On the Edge
of the Wild

On the Edge *of the* Wild

Audrey Tournay

MUSKOKA PINE PRESS

Copyright © Audrey Tournay, 2008

Published by
Muskoka Pine Press
Rosseau, Ontario
P0C 1J0

Available from
Aspen Valley Wildlife Sanctuary
1116 Crawford Street
Rosseau, Ontario P0C 1J0
Phone: 705 732 6368
Fax: 705 732 1929
www.aspenvalleywildlifesanctuary.com

Library and Archives Canada Cataloguing in Publication

Tournay, Audrey, 1930–
On the edge of the wild / Audrey Tournay.

ISBN 978-0-9811103-0-1

1. Wildlife rescue — Ontario — Parry Sound Region. 2. Wildlife rehabilitation — Ontario — Parry Sound Region. 3. Tournay, Audrey, 1930–. 4. Aspen Valley Wildlife Sanctuary (Ont.). 5. Animals — Anecdotes. I. Title.

SF996.45.T677 2008   639.9'50971315   C2008-906788-6

Design by Gillian Stead
Printed in Canada by Friesens

Dedicated to

the memory of

Joey and Hamish,

faithful canine companions.

Audrey and Nim-keas-quay.

Contents

	Prologue	9
I	**Dogs** — Assistant Caregivers of Wildlife	
	1 Good Dogs and Good Deeds	13
II	**Small Animals**	
	2 And God Created Skunks	25
	3 All Those City Raccoons	39
	4 Moses and the Sixth Commandment	57
	5 A Bittersweet Story of Two Beavers	68
III	**Coyote**	
	6 The Remarkably Clever Coyote	75
IV	**Wolves**	
	7 In the Beginning — Learning About Wolves	83
	8 The Winter of the Wolves	92
	9 Amarook, the Wolf From Way Up North	106
	10 The Christmas Wolves	111
	11 Once Upon A Time There Was a Wolf	113
	12 A Cello in the Woods	115
V	**Bears**	
	13 The Story of the Three Bears	131
	14 Shaman Bear	149
	15 Nim-keas-quay	154
	16 The Story of Fuzz, the Bear	160
	17 An Overbearing Situation	174
	18 Sydney Bear's Springtime Release	188
VI	**Larger Animals**	
	19 Days of the First Deer	201
	20 Chewy the Moose	217
	Epilogue	235
	Aspen Valley Wildlife Sanctuary	237

Luna

Prologue

I unfolded the blanket in the back of my car. The veterinarian looked down at the young coyote, curled there, unmoving, red, raw with mange. Its eyes were closed. It was cold.

The veterinarian said, "It's almost dead. I don't know if I can help it."

Four months later I stood with the coyote, far out at the edge of a remote wilderness. His nose was pressed against the door of the cage; his legs were braced for running. His eyes were fixed, not on me, but on the field and trees. And I opened the door. He ran. He ran gloriously free. He did not hesitate. He did not look back. He was back where he belonged. Free!

That experience — the experience of seeing an animal that had come to the Sanctuary orphaned or injured, returned to the wild — is a privilege which I have experienced time and time again. To see a raccoon, which as a tiny kit, eyes closed, orphaned — to see that raccoon, adult, climb a tree in some place far away from humans, to see his excitement — to see a fox, leg fractured by a fast car on the highway, run across a far meadow and vanish into the tall grass — to see an owl, wing broken in some unknown way; be able to spread its wings and fly away into the nighttime shadows. All of this is a privilege beyond explanation.

"Privilege" is the best word I know to describe the experiences I have had and that I am sharing with you in the stories which follow. Being with the orphaned and injured animals which come to us, watching them through the long process toward maturity or recovery,

I have learned a deep respect for the moose or the mouse . . . all creatures great and small. I have learned to respect the wisdom which they have; I have seen them show compassion and humour. I have seen them be what, to our way of thinking, would be called "cruel". I know that they have depths in their lives which no researcher, biologist or writer can ever fully understand.

To these we owe freedom.

Inevitably, some animals come to the Sanctuary which, because they have been so possessed by humans (well meaning or not), cannot go free. Wolf/dogs, coy/dogs, blind raccoons, descented skunks — such animals as these would have no chance of survival in the wild. To them we owe the best life possible — huge enclosures, as nearly as possible like the wild places where they should be living — the food they eat, also as nearly as possible like the food they could find for themselves in the wild — and, when possible, the companionship of their own kind. These animals give us a chance to watch, to begin to understand them, to appreciate their intelligence, their personalities — their friendship. We must respect their right to privacy; we know they will not fit any pattern we think should be theirs. A wolf/dog is unpredictable — torn between two worlds — as is a coy/dog. No bear is ever truly tame. Nor is a raccoon or a skunk.

And, as a "by the way", people need to be careful to try not to read into animals the thinking and actions of humans. Anthropomorphism is a dangerous temptation and becomes more and more difficult to analyze or recognize!

So I wish to share these stories — stories of animals caught between two worlds. I have walked in the woods with bears. I have been swimming in the ponds with beavers. I have sat in sunshine on a rock in the woods while a deer grazed nearby and then came and lay down beside me. . . . I have had privilege beyond imagination.

Prologue

Audrey with Blackberry, the porcupine.

I also want to express my deep appreciation for the many humans who make the care of wildlife animals and the work of the Aspen Valley Wildlife Sanctuary possible — the Sanctuary staff, especially Tony Grant and Janet Longhurst, the Board of Directors, and the multitude of volunteers and supporters. I am also deeply indebted to those who have made it possible for me to share, in this book, some of my wonderful experiences with wildlife, especially John Denison for his advice and guidance regarding the production and printing of the material. My sincere thanks to all of you — my friends and supporters in my adventures with the creatures who live on the edge of the wild.

Kelly

Laddy

DOGS

I

ASSISTANT CAREGIVERS OF WILDLIFE

1 – Good Dogs and Good Deeds

Early one morning I was driving along the highway on my way to teach school, not thinking especially about dogs, nor really any other creatures, letting my mind relax and wander. One needs some relaxation before confronting the world of classrooms and timetables and demanding students. At least I did! Then I saw a Collie dog standing on the top of a small rock-cut, looking down at the passing traffic. A slight breeze touched his long hair and his tail . . . I thought: "That's a beautiful dog." And I thought about him the rest of the way to school.

I didn't need another dog. Three black almost-Labrador retrievers were living with me. The four of us filled my tiny house — the four of us and any wild creature who happened to need a place to stay for awhile.

Still, I kept seeing that beautiful Collie standing in the wind on the rim of the rock-cut.

In the staff room I met a teacher who I realized lived across the highway from that particular slice of rocks. I said: "I saw a lovely Collie standing near your place this morning — is it yours?"

She sighed in some exasperation. "No. A stray, I guess. He keeps coming to the door but we don't want a dog in the house. Do you want him?"

"No." The four of us filled that little house.

She shrugged her shoulders. "We're taking him to the Humane Society tonight."

That, I knew, was a sentence of death. Because hunting season was just over, the Society was swarming with lost and abandoned hunting dogs — far too many dogs ever to place in good homes. Because they could locate and even herd deer, Collies were sometimes used in the hunt. Not often. But sometimes.

I had work to do. Students to teach. I tried to put the dog out of my head. But, when the school day was finished and I was driving home, I started to remember. Out at his farm, years and years and years ago, my Uncle Allan had had a Collie dog named Laddy. Uncle Allan would send me, a very young city child, out to the pasture to "fetch the cows home". Twelve cows, I remember.

Always he made sure that Laddy went with me. I realize now that apart from opening and closing the pasture gate, Laddy could have done the entire task without me. Laddy was a black and white and brown Border Collie. Still, back in those days, like most every child, I kept a scrap book of pictures cut out of the Toronto Star Weekly. One picture I loved was a close-up of a Collie who looked just like Laddy and I thought, "Some day I will have a dog just like that." Instead, I had three black dogs whom I loved. With me, we filled my little house.

The next day at school the teacher said to me: "We took that dog into the Humane Society."

The manager of the Humane Society, Carol, was a friend of mine. As I walked around the tables of students in my art class, helping them with this and that, my mind was still on the dog. I thought: "Carol won't mind if I just phone to find out if the dog is likely to get a home . . ."

So, that noon hour, I found myself dialing the Humane Society number.

I heard myself say: "Don't put him down. If nobody else wants him, I'll take him." I couldn't believe I was actually saying it. That had not been my intention! And I should have known exactly what

Carol's reaction would be; she had dozens of dogs to find homes for. Laddy would be mine. On the way home from school I stopped to visit him.

Carol led him from his backroom cage out to the visiting area. He didn't look like either Uncle Allan's dog or the dog in the Toronto Star Weekly. I rubbed his ears. He looked at me. His eyes were not the big, soft brown appealing eyes of a Border Collie. But they were intelligent and very wise. So were Carol's.

I said: "Only if no one else takes him."

She smiled. Nodded. Returned him to his cage.

At the end of the week, I took him home.

His ancestry was primarily Border Collie but from the shape of his head and the eyes, I suspect that somewhere in his background was a German Shepherd. However, the Collie thought-patterns were certainly there. Not only did he come when he was called, he came when he was needed.

Across the laneway, between the house and the barn was a large bird cage where I was raising young, orphaned ravens. Ravens are generally a noisy lot and certainly the cleverest of birds so I was concerned when I heard them making a considerable racket.

Laddy was also. He trotted over to the cage and around to the front of it and came loping back to me, barking with alarm. He stood in front of me, braced to have me follow him. Which I did — around to the front of the cage in time to see that a large raccoon had managed to burrow into the cage and was most certainly meditating about raven for dinner.

As soon as Laddy saw that I had the situation under control — the raccoon out and me busily mending the entrance hole, he lost interest in ravens and raccoons. His part was finished.

For several years, our neighbour, about a kilometer up the road, decided to keep sheep. He had only about twenty of them but enough that, when they found a weak spot in their fence, they could come flocking down the road and raise havoc in amongst our enclosures. Laddy understood about sheep and somehow he knew exactly where those sheep were supposed to be — how he knew I do not understand. The farm was up a hill and around a bend, through woods and rocks and down another hill. However, Laddy knew.

At the first sound of approaching sheep, Laddy was on his feet and whipping up the road. He not only herded them away from our place, but he took them all the way home.

Now, looking over my huge collection of photograph albums, I see pictures of Laddy sharing his food bowl with raccoons, playing with Bucky, the deer, in his pen, examining and lying down beside a litter of baby skunks. My favourite pictures are of Laddy with a beaver kit. In one picture he is lying down with the kit between his front paws and his nose touching the kit's uplifted nose. In another, his paw is around the kit's plump backside.

Often I have heard the objection: "You shouldn't have dogs at a wildlife Sanctuary. The animals will get used to dogs and that will be very dangerous for them when they are set free!" However, wild creatures are somewhat more clever than that. We humans can tell one dog from another — we know which dogs are ours, which dogs belong to the neighbourhood, which dogs are strays or dangerous. Wild animals are at least as knowing as we are, if not more so. Laddy and Bucky, a deer, who was a permanent resident with us, were good friends. They romped together. They slept side by side in the sunshine. Yet, if a strange dog arrived, the deer was nowhere to be seen. Time and again I have observed the same reaction. And our dogs have made a very real contribution to the rehabilitation of wild things.

During the course of many years, I have raised and loved fifteen dogs. My first dog was a pure bred German Shepherd whose ancestry was

traceable through the show rings of Germany and London, England. Joey was the only dog I have ever purchased. She is the only pure bred dog I have ever owned. We lived in St.Catharines, Ontario. I had never thought much about wildlife. I had never even considered moving to Muskoka. Joey and I went for long walks together along the Bruce Trail which traces its way through the Niagara Escarpment. And then (foreshadowing?), a black stray arrived on my doorstep. After several days of stubbornly refusing to let him in, one night during a bad blizzard, I opened the door to him, and subsequently, to the other thirteen dogs who have been my friends.

At first I considered Angus to be a black Labrador. The members of the St.Catharines Kennel Club (who welcomed my lovely purebred Shepherd), acknowledged he might have some Labrador blood but he was actually a black mutt. I argued that even Labs were allowed to have some white on their chests.

"Yes", came the reply, "but not a whole damned constellation!"

What Labrador blood he did have was evident in his willingness to retrieve. One day in the early spring, when a thin layer of ice still skimmed the water, the dogs and I were walking beside the Twelve Mile Creek. From a branch of a large maple which overhung the water, a small black squirrel tumbled, broke through the thin ice, and struggled to get out. It couldn't. Without any hesitation, Angus leapt out, breaking the ice as he swam, seized the squirrel in his mouth and brought it in laying it gently at my feet. And, tail wagging, looked up to me for approval. I told him what an excellent dog he was.

That rescue began his career of retrieving wildlife. Squirrels often. Baby rabbits sometimes. And, occasionally, skunk kits. I do not know how much altruism was involved but, Angus was a retriever and he was intent on pleasing me. He never harmed or killed.

He was eventually killed by a speeding car. Humans can be dangerous.

Later, after I had moved to Muskoka, two more black mutts came to live with me. Dr. Alan Christie, the veterinarian in Parry Sound, who was such a help at the beginning of the Sanctuary, had found a box of black puppies abandoned beside highway 69. I inherited the last two — Black Lab and sometimes I suspected coyote, but only their mother would know for sure. Kate and Abby were siblings, but in character very different. Abby was smaller, quicker, intent on exploring, understanding and dominating the entire world. Kate was larger, a quiet, dignified, responsible lady.

Abbys' best friend was a large, domestic Mallard duck named Archie. She and Archie would roam around the barnyard together, stroll between enclosures, Abby with her nose to the ground, Archie clucking in continuous conversation. In the warm sunshine they would sleep together, a tangled heap of dog and duck. When we wondered where they were, we could follow their sandy footprints up the road — dog and duck — side by side.

Kate had a more serious outlook on life. Far more accurately than did her sister, Kate understood the responsibilities of life at an animal Sanctuary. While Abby was out romping with Archie, Kate was doing all that was within her ability to supply the care needed by small orphaned raccoon kits. Spring always was, and still is, the season when the orphaned young almost overwhelm the human caregivers — small creatures, eyes closed, demanding food and warmth, needing the feel of the mother's body close, needing her care. Perhaps the mothering instinct in Kate was strong but, for whatever reason, she did all within her power to care for little raccoons. She stayed near their boxes. As I finished bottle feeding each kit, I would put it down beside her. Kate would groom it, licking the little belly just as its mother would have done, and then tuck it as close to herself as she could. Though she was a spayed female and could produce no milk, Kate allowed the kits to soothe themselves by suckling on her.

Once, perhaps somewhat jealous of the credit given to Kate for her concern, Abby allowed raccoon kits to try to nurse from her. The experiment lasted, at the most, half a minute. Abby stood up and shook the kits away. No mother instinct here — a committed spinster!

Sadie and Lucy knew little about responsibility but they did know how to enjoy life! Lucy was mostly Australian Sheep Dog, grey and white with blue eyes. Sadie, a stray whom I had found as a pup, barking under my car in the laneway — an obviously deliberate abandonment. She was black and white, maybe part terrier — maybe hound — maybe something it would be a challenge even to God to figure out. During the day the two dogs shared a large enclosure around a good sized round beaver pond where, generally, a beaver or so were also living. Cattails and lily pads made a good environment for frogs and little fish and the occasional turtle. While the fish and turtles were unmolested, the dogs passed a good part of the day chasing frogs — though I do not recall that they ever managed to actually catch one.

At least not until a year when we had no beavers. Beavers, being vegetarians, never bothered the frogs; bear cubs are a different story altogether. I had been raising a bear cub named Neech-ka since he had been no larger than a black squirrel. Now, for part of every day, when I could keep an eye on him, he was allowed freedom. He began by walking along the top of the fence which surrounded the beaver pond, and tantalizing the dogs. When he saw that they were beginning to ignore him and return to the chasing of frogs, he climbed down and joined them. So the three, the two dogs and the bear cub, would circle the pond, the frogs would jump out into the safety of the water . . . .

Until Neech-ka seemed to realize that no bear ever survived without actually eating the frogs, life was going to be more serious for him. He abandoned the dogs at the water's edge and swam out into the pond. Then, when the dogs chased the frogs and the frogs jumped in, the cub was ready for them. He ate them.

I rather think that Sadie and Lucy felt that Neech-ka was ruining their game. The frogs certainly did!

Kate's attachment to helping with the raccoon kits could have been explained by a strong mothering instinct. Hamish was a big, neutered male Labrador.

Four, it seems, is a very common number for a litter of raccoons. I was raising my first litter (how long ago that seems!), so long ago that I still named every creature who came to us. These were Pippin, Merry, Sarah and Little. Little was a runt and I didn't expect her to live. (She did, and stayed freely around the Sanctuary for several years. She also managed to kill an entire flock of chickens — with the best will in the world! However, that was all in the future). As I bottle-fed the raccoons, Hamish went about the business of his life, being a male dog with the responsibilities of walking the perimeters of the land and announcing the arrival of any humans — certainly not helping with babies. However, as the raccoons grew, his interest deepened and for a reason I did not understand, he and Merry became special friends. All their first winter the raccoons spent some place in the bush. We did not see them. When spring finally came, Pippin and Merry and Little returned. Hamish and Merry resumed their friendship.

Merry always waited under the evergreens by the side of the laneway. Around six o'clock I would put a bowl of kibble on the porch when Hamish would come for his dinner. When he came, Merry would come, too, go under or around him and eat with him. He would shift to allow her room. When they had both finished, Merry trotted down the lane and away into the bush across the road. Hamish went about whatever business he had in mind until the time came to come into the house, crawl up on my bed, and go to sleep.

One evening Merry did not come. Nor the next. Or the next. At first Hamish seemed merely puzzled. And then, agitated. Ignoring the bowl of kibble, he poked me with his nose. He started down the lane, stopped, and looked back at me. When I did not respond, he came back and poked me again and started down the lane, once more looking back and almost commanding,

"Come!" I followed.

Across the road, through a shallow creek, up the hill and into the bush. Hamish kept looking back to make sure I was following. Finally, he stopped by a newly fallen tree. Merry was pinned beneath it. She was dead. But her four kits were not.

When I picked them up, Hamish whined and bounded back toward the house, stopping only to make certain I was still following with the kits. While I fed them, he stayed close beside me. He slept beside them. They were his. And Merry's.

One of the kits, Freddy, did not thrive. All summer I worked hard to keep him alive. Hamish guarded him — and continued guarding all through the next winter, while Freddy slowly gathered strength. Other raccoons Hamish blissfully ignored, but until the day when Freddy went free, Hamish never abandoned him.

These have been stories of dogs doing good things. Some theologian whose name I forget (I am excellent at forgetting the names of theologians!) , said that in order to be capable of being truly good, one must also be capable of being evil. I know of one grievous sin Laddy committed: he did kill a woodchuck.

Even Mother Theresa, I have read, had days of darkness and doubt, a fact which I believe, adds to her greatness as a saint. Goodness did not come to her naturally — she had a choice. So did these dogs. So did Laddy. And if someone feels that comparing Laddy to Mother Theresa is an affront to her, I am sure that she, herself, would not think so. Neither she nor God. And I hope that in the great "Whatever-Comes-After", Theresa has found eternal light and Laddy has apologized to the woodchuck.

In a famous English painting, a Collie is depicted standing in the midst of a raging blizzard. He has found a stray lamb. His nose is raised in the air, as he howls for help. Laddy had that same intelligent nose and the same concern.

In the spring, when the highways are streaming with the cars of city folk seeking the country quiet, the road-kill count is high — raccoons and skunks, foxes and porcupines, bear cubs and deer — many of whom are nursing mothers. If the babies are not found, they die. Because fawns are born with no scent, the mother doe is able to leave them for long periods of time while she grazes. Well hidden and scentless, they are relatively safe from predators. If the doe becomes a highway statistic, the fawn has no hope of survival.

That is when Laddy's abilities were essential. Exactly how he could find an odourless fawn, I cannot explain. But he did. He was good at finding orphaned raccoon kits, too — or, at least, finding the tree which they had climbed up for refuge. But even a talented Collie cannot be expected to climb a tree!

One of Laddy's best friends was a coyote. Coyotes are extremely intelligent, extremely independent and they have a wicked sense of humour. Two coyote pups, eyes closed, small and as helpless as coyotes ever are, were in a small dark box in the house next to the wood stove. Every few hours I had to bottle feed them. I cleaned and stroked them as their mother would have done, but I had the impression that, even though they were so young, they knew I was a mere substitute caregiver. However, until they were weaned, they had to be content with the care I could give them. Once they could eat on their own, they did not want to be touched. And, since they were to go wild, that independence was to be encouraged.

A neighbour's daughter named them Thunder and Lightning.

Even Laddy could not help me build enclosures, and since no one else was available, I set about adapting an existing one which had, at various times, held raccoons and foxes and skunks and the occasional beaver. I strengthened the surrounding fencing and the posts. I spread chicken wire across the earth floor so the coyotes could not dig out. I stretched chicken wire across the top so that they could not climb out. I put in a dog house for shelter and evergreen branches for privacy. The pups, living there in what seemed obvious acceptance, grew. Through the fence, Laddy made friends with them.

One September morning, when I went out to feed the pups, I discovered a hole in the chicken wire and the coyotes were gone. Laddy, not quite grinning, looked at me and wagged his tail slowly.

Though I knew that they were big enough to care for themselves, I was exasperated. I had wanted to release them far from any human contact. Humans are dangerous to coyotes. I suppose I should have read more into the sparkle in Laddy's eyes; he knew exactly where, in the surrounding meadows and woods, those coyotes were. In the days that followed, I would see him up in the meadow, where the rocks rimmed the edge of a stand of huge pines, sleeping in the sunshine with the coyotes. They never came near the house. Laddy never went off into the bush with them.

I knew I would never be able to recapture the coyotes, so I didn't try. They never became a problem to humans. They never became a problem to the sheep up the road.

Eventually, Thunder went off with the wild coyotes and I seldom saw him. Lightning came regularly to visit — not me, Laddy. I would see him standing up on the rocks, waiting. Laddy would lope up the meadow and the two of them, side by side, would vanish somewhere. A couple of hours later I would see them, on the rocks again, sleeping in the sunshine. After awhile, Laddy would come down to the house and Lightning would disappear. That friendship continued for several years until, something having happened that we know nothing about, Thunder no longer came.

Sometimes wild animals which cannot be released come to the Sanctuary — declawed and defanged bears, wolf-dog crosses, foxes raised in apartments, descented skunks. And raccoons, too, often blind. These animals usually come with names. I think Mr. Disney bears a terrible responsibility here. Fawns are almost inevitably called: "Bambi", "Buck" or "Doe" — it does not matter. Coyotes are "Wiley", Skunks are "Pepe lePew" or "Flower". Over the years we have had dozens of raccoons named, "Bandit".

One particular blind raccoon named Bandit had been raised in Toronto and was brought to us only because the people who had had him as a pet were moving away and could not take him. Through the years they had been good to him, though raccoons do not make good pets. (Sexually mature, they will attack and bite. Their bite can be deep and painful). However, Bandit remained a good raccoon. I had some help with the building this time; he had a large enclosure, a shed, a pool and an evergreen tree to climb. For several years he was quite content, and then, one night, he managed to make a small hole under the fence. He was gone.

Where, in four hundred acres of trees and meadow, hills and wetland, does one begin looking for a raccoon? I looked. All day I looked. Laddy ranged the fields. No Bandit.

For two long days and nights — no Bandit. I was not certain how long a blind raccoon could survive on his own in a world of owls and coyotes and wolves.

Early the third morning, I was sitting on the porch, cup of coffee in hand, wondering what I could do. I saw Laddy, away at the back of the meadow, where a rocky cliff drops down into the edge of the wetlands, his head down, moving slowly.

When I called him, he ignored me.

Slowly, up through the long grass of the meadow, he came.

And I saw, as he came near, that he was herding, gently and slowly, one blind raccoon. He had found Bandit and brought him home.

Dogs can be very, very good.

SMALL ANIMALS II

2 – And God Created Skunks

I named the skunk Charles, after my father, who was a clergyman. The skunk, so strikingly black and white, suggested clergy to me; further, I thought that, if the skunk, large, overweight and slightly pompous, had any theological inclinations he might have been a bishop or even an archbishop. Since my father was neither a Roman Catholic nor an Anglican (nor anything else that might lead to a bishopric of any sort), I'm not certain if the clerical connection was justified. Still I named the skunk, Charles. When it comes to current theology, I'm quite certain that the skunk and I would have understood each other very well.

Perhaps the connection was centred in his dignity! He was very round — actually he resembled a doughnut on four legs — and, with his plumed tail like an umbrella over his back, he moved with a measured, calculated stateliness, as though he led a solemn procession. All this was, I am quite sure, engendered by the fact that he grew up bored.

The first eight years of his life were spent in a cage in an apartment in Montreal. A dog and a cat, also enclosed in the apartment, were his only contact with life — and how much they cared about him is a matter of speculation only. Certainly the humans who lived in the apartment cared very little. Eventually they grew tired of the animals, packed them into their car and took them down to the Montreal SPCA to be killed. However, a friend of ours who does rehabilitation work in that city happened to be visiting the SPCA. Though she could not do anything for the dog and cat (they died), she could lay claim to the skunk. And she did.

She phoned our Sanctuary. "Have I got something for you!" Since her Urban Animal Advocates Centre has presented us with foxes, bears and coyotes, and since I knew she cared for larger mammals such as seal cubs, which had drifted down the St. Lawrence, I held my breath and said: "Okay — what now?"

"A skunk. Who has been descented." I knew she had no facilities for a non-releasable, life-long resident skunk and besides, a visit from Harriet is always exciting. I said: "Yes, When?"

The next day she put Charles into a kennel, and the kennel into her van, and drove the long highways from Montreal, through Algonquin Park, to Rosseau. Charles had matured in the throbbing noise of a large city. He would spend the rest of his life in the quiet of a wilderness valley. Harriet believed that the long trip for a single skunk was worth every mile.

How does one induce a lethargic, bored skunk to take an interest in life? How to coax a fat skunk to become active? First, instead of a small cage, he was given a large enclosure where he had to walk at least ten feet to get his food, and another ten back again for a drink of water. That seemed worth a try, but it didn't work. He simply wandered in an unhurried manner through the straw, under the evergreen branches, sat down wherever he felt inclined and went to sleep.

During that time, a beautiful, young, female skunk, also descented, was brought to the Sanctuary. One instinct is basic to all living things. Surely she would be the answer to an elderly skunk's dreams. We filled his bowl with food, gave him fresh, clean water and a new hollow log, big enough for two. We put Pia, young and slender and beautifully marked into the pen with him.

She beat him up!

As winter approached, we moved Charles to a more sheltered pen beside the barn. Neighbours in the adjoining enclosure were three splendid young skunks whom we had raised from tiny, orphaned babies and who were awaiting spring release. Charles actually noticed them. He stood, eyes opened, nose twitching, watching. He moved toward the fence. They were nose-to-nose through the fence — friendly? I remembered Pia. I waited a few days, and, then I put Charles into their pen. Instant ecstacy! They ate together. They slept together. They played. Charles played. He lost weight. He took care of himself, cleaning and grooming. Instead of a pale yellow, his stripes became gleaming white.

When the winter grew very cold he went into the kennel with them and they pulled the straw into the doorway blocking the cold out and the warmth in, and, while the snow whirled across the valley, they slept. For weeks. Spring came — green, growing things, sunshine, romance. Time for youth. Those three skunks beat him up!

The three were released. Though alone in the pen, Charles did have a gleam in his eyes. He had barely settled into his bachelor existence when a skunk, named Spectacular, arrived home.

The previous season I had raised Spectacular from a tiny kit, and, because she was such a remarkably beautiful skunk, given her her name. The stripe up her nose was wide. Her white cap covered her head, curled around her ears, and joined beneath her chin. A single white stripe spread like a cape over her shoulders and divided into two broad stripes down to her tail, which was full and gleaming white. We had released her some distance from the Sanctuary. She had travelled — across a river, through a large wetland, along a highway and into the yard of a mechanic whose wife wanted a pet skunk. She phoned me and told me about the remarkably lovely skunk hiding under a rock, out behind the garage. Would I help her catch it?

I went out. The description was too co-incidental for the skunk to be anyone other than Spectacular. I caught her, explained that she was not descented and that descenting was illegal in Ontario, and took Spectacular home with me. We released her right at the Sanctuary; there we could watch her.

When winter returned we were really crowded for space. I put a second log in Charles's enclosure. But Charles had learned. All winter he kept his distance.

In the spring, he beat her up!

Charles lived with us for many years. One winter, when he was an old, old skunk, he simply went to sleep and did not awaken. I don't really know if he and God discuss theology in heaven.

If I were making a list of my favourite things, I might work through raindrops on roses and whiskers on kittens but would definitely begin with my dogs and the beavers whom I have been privileged to know, and assorted bears and coyotes and wolves and raccoons, (maybe even some humans) — but, without any question, away up at the top of the list would be skunks. Some of my best friends have been that small, black and white, much maligned mammal; not descented — free when it could be free — but always delightful. Though I have been told dozens of stories about someone who knew someone who had a skunk for a pet and "it was just like a cat", few of the stories stand up to investigation. Like all wild creatures, as soon as they are mature, they want to be free.

When they are born, skunks weigh only a few ounces. Once I held a new born skunk in the palm of my hand, tiny, hairless, though the skin bore the black and white markings, eyes tightly closed and, unfortunately for this one, very cold. It had been found lying on the cement floor in the barn. We never found out how it came to be there. Though we searched, we never found a mother. Perhaps a wild skunk, very near to the time of birth, had been badly frightened somewhere in the barn — who knows? I tucked this tiny being under my sweater to keep it warm while I found the heating pad and blankets. I left it to warm while I found the eye dropper and Esbilac to feed it. So very, very small and new born and wonderful. But too cold, too long.

Though skunks in the same litter are usually very similar, they are never identical. Skunks from different litters may vary widely — two stripes, no stripes, thin stripes, wide stripes, stripes with little flares at the shoulders or flanks, little black buttons in the stripes at the neck or in the nose stripe. Once I had a skunk with a complete white back. But because of the variations, each individual skunk is readily identifiable.

I remember the first time I ever saw a black skunk. That happened long before I ever dreamed of becoming involved with wildlife. It seems like years and years ago when I was in my first year of teaching. Then, away back in the 1950's, teaching was very different from what it is today. Just to illustrate. I was a first year teacher, completely inexperienced. However, I was given as my first class, a grade nine, all boys, all repeaters — all forty-seven of them! I understand that, during the long years which have followed, six of these students have ended up in a penitentiary — for which I take no responsibility.

Back to skunks.

I lived in a second story apartment at the back of a house on a residential street. The apartment was reached by an outside staircase. Arriving home one evening, just at dusk, I saw a movement on the ground under the staircase. I stopped. Looked. A skunk?

"No", I told myself. "It doesn't have any stripes".

Later, when I told a friend about it, she said: "It was a cat. Skunks have white stripes. It was a cat, silly."

Who was I to argue — away back then?

However, most skunks do have two white stripes. Quite often I have heard people express surprise at that. "Two?. I thought they had only one!" Blame the mistake on movies and comic books. Though I have raised dozens and dozens and dozens of skunks, I have never encountered a single striper.

Once, Charles and I were at a school where we had been invited to talk to children about animals. A grade six child, hoping to be regarded as an expert (her father was a trapper, and who would know better?), looked at Charles wisely and emphatically informed her classmates:

"That shouldn't be called 'Charles'. It's a girl!"

"No". I said reasonably, about to turn Charles over so that his sex organs could be identified. "Look."

"It's a girl!", she nevertheless insisted. "My Dad told me that girl skunks have two stripes and boy skunks have one!"

Never argue with a parent.

Snowman was a pure white skunk, an albino. Bill, a friend in Toronto, spent a good part of his life rescuing exotic, abused animals. He had several panthers who had had careers as companions for strippers. He had a row of extremely noisy macaws and several monkeys who had once been cute little fellows for sale in pet shops. But even monkeys tend to grow up!

When Bill acquired an albino skunk, he phoned me. "Pure white", he told me, "and descented. I haven't room. You need a white skunk".

Bill had managed to take him from a gang of young men who refused to say where they had found him. When they injected themselves with drugs, which appeared to be their major entertainment, they also injected the skunk. I suppose watching an albino skunk high on drugs was, in itself, some sort of "high" for them.

Though Snowman had his own enclosure at the Sanctuary, and was given good food and water, his own log and evergreens for privacy, he never became a normal skunk. Though he ate, slept in the log and sniffed around his pen, vaguely interested, he was always slow, his approach to life entirely without curiosity and initiative. He did not seem to notice if he were alone, or being picked up and stroked. His lethargy was not due to his albino condition (we have raised other albino creatures who were perfectly normal), but, I am sure, it was because of that first drugged year. We gave him the best life he could have.

A much more tragic story than Snowman's occurred a few years after his death. Spring, and once again a young skunk was brought to us, another skunk whose owners had thought giving a skunk injections of drugs was a hilarious pastime. Her reaction was entirely different from that of Snowman; she was continually frantic. When she ate, she simply grabbed her food and kept moving — when she was asleep, her legs were still moving and her body was never still. At first we hoped that the effects would gradually fade; they didn't. One night she managed to tear her way out of her cage.

We searched for her — through the woods, around the rocks, along the edges of the pond, through the wetlands. Once, a few weeks later, we saw her briefly. She was still frantic. A day later we found her dead. Her nipples were swollen. She had been about to give birth.

Naming a skunk is a real responsibility. Cartoon skunks have dismal names — usually after flowers or French fragrances. And so I feel I do owe an actual apology to one of the most beautiful skunks I ever knew. I named her Blossom.

Very late in the season, Blossom had been found wandering all by herself down a highway very near to a golf course; she should not have been alone. She should have had a mother or at least one sibling — but golf courses and their pesticides are very fatal to wildlife. Blossom was alone and somewhat frightened. When she was brought to the Sanctuary, she was a small bundle of absolutely beautiful black and white, but frightened and very hungry. All of the skunks we had raised that year had been released; this skunk was too small to withstand the cold of a winter in the wild. Against all better judgement, but having very little alternative, I took her into the house to stay with me, with the dogs, the cats and the beaver. They accepted her. She accepted them. No problems.

That winter, traveling in a small kennel, Blossom accompanied me to the schools where we were frequently invited to talk to the children. They never knew what animal I was bringing with me and I never told them until I was in the classroom and talking to the children. The kennel would sit, undisturbed and dark while we talked about wolves and bears and beavers and raccoons. Then, as a climax to the talk, I would turn to the kennel, open it and whatever animal was with me would walk out and meet the children. Only when they saw Blossom walk out — and not until then — would the children's hands fly to their noses and they would squeal about the awful smell. The smell! Suddenly unbearable!

"Ooooh — she stinks!"

"She stinks! She stinks!"

The teacher would stand rigidly silent.

I would wait. When they had settled I would remind them that Blossom had been in the room with us all along. They had not smelled her until they had seen her. Gradually, the hands would come away from the noses. Tentatively, curiosity would take over. And then, pleasure and delight as they began to see her, beautiful and gentle and as curious about them as they were about her.

Usually, the teacher was the last to succumb. "She is descented, isn't she?"

No. She was not descented. Some day she would go free. That was another entire addition to the classroom talk.

Blossom became a television star. A television crew from the BBC in England had asked to come to the Sanctuary to make a documentary about beavers. That was fine and the documentary they produced was fine as well. It began with several actual shots of Grey Owl and his beavers, then moved to a campfire amongst the pines and rocks on the cliff beside our valley. There, as the sun was going down, a group of Ojibway children listened attentively as one of their Elders told them stories and legends about the beaver people. The next day they came to the Sanctuary to meet a real beaver. Quibble, our resident beaver, swam in the pond for them, dutifully carried the aspen branches we gave to him, packed them into the dam and allowed himself, sitting on my knee, to be carefully examined. They filmed his front feet, like hands, his back feet, webbed like a duck's, his orange teeth and his big flat tail. In her enclosure, Blossom completely ignored all the activity. It was more difficult for the television crew to ignore someone as beautiful as Blossom. They asked if they could stay another day to do a documentary about skunks.

Blossom was small for a skunk, and quite round. Her stripes were wide and very white. Her tail pure white, was huge and curled over her back, or following her, was always moving as though brushed by a gentle breeze. She trusted people.

For the long hours of the day the crew handled her, posed her in appropriate places, followed her while, nose down, she hunted for grubs and took her to raspberry bushes where she obligingly ate. They got wonderful pictures — Blossom thrusting herself through a field of daisies, mirrored at the edge of the pond where she hunted, peering out of a hollow log. Not once, during that long, long day did she threaten to spray.

A couple of months later we received copies of the documentaries which had been made. The beaver documentary was excellent; Quibble was at his handsome best. Blossom's pictures were good, too. Until, suddenly, as the film drew toward its end, Blossom vanished. A cartoon skunk, marked exactly like her, took over. The cartoon skunk was frightened by things Blossom would never have noticed — but it was frightened. It stamped! It threatened! It sprayed! Again, at every opportunity, it sprayed. They had had to get a cartoon to do it. I felt that Blossom had been betrayed.

The reason for the very existence of the Sanctuary is that the creatures who come to us, if it is in the realm of possibility, shall be returned to their free lives. They will live as they are meant to live. Usually the animal is taken to a suitable habitat, carefully chosen, far away from dangerous human beings. But sometimes — just sometimes — an exception is made. At the Sanctuary we have a few hundred acres of wilderness. We are surrounded by much, much more. Thus, an animal like Blossom, well . . . one night in the early spring, when I knew that new food was plentiful, I went out and simply opened the door of her pen. And whispered goodbye. Sometime during the night, she slipped away into the darkness.

I thought about her often and wondered: was she hunting well? eating properly? had she found a safe place to live? a mate? Though her enclosure remained empty and the door open, she did not return.

That spring passed and a hot, hot summer. Baby skunks with broad stripes, narrow stripes, no stripes, arrived at the Sanctuary. Three dozen of them, in small enclosures scattered around the barn and back into the meadow. They shared our time with all the other arriving creatures — bottle fed with the raccoons and beavers and foxes and coyotes. Still, I sometimes thought about Blossom, away from all our busyness, somewhere in the free places, moving unobtrusively about her own life.

Summer, and then the autumn, when the brilliant colour of the leaves made the entire world a place of glory; a time when, for two long months, the bears are hunted. We were overwhelmed with orphaned cubs. (Never believe, as we are so often told, that female bears are not killed — that the hunters are careful. Never, never believe it. If you are tempted to, visit the Sanctuary during the bear hunt and after it when the winter is cold and the orphaned cubs are found, starving). Then, one day after long, long hours of work caring for bear cubs and other creatures in the Sanctuary, I was standing on my porch, being still for a moment. Inside the house, the dogs were waiting to be fed. On the porch, the cats were staring at me reproachfully. I was late with their meal, too. And then I saw a skunk coming slowly up the lane, unafraid, toward us. Not just any skunk.

Blossom.

Out of the wilderness, Blossom had chosen to come home. Perhaps she had something proudly to share with me. Behind her, with wide white stripes and floating white tails, were two little Blossoms.

I held out my hand to her. For a brief moment she touched her nose to my finger. Then she took her children under the porch. I brought out a bowl of cat kibble.

Then I fed the rest of my family.

Blossom and her kits moved into a small kennel which had developed under the floor of one of the bear enclosures. For that winter the skunks lived underneath the hibernating bear cubs. When the blown snow lay along the aisle between the enclosures, I would see it patterned with tiny footprints and know that Blossom and her family had been looking for food in the bear pens. When they found nothing they would invade the pen where Bandit, a blind raccoon, was curled up asleep. They would eat his dinner.

When spring came once again, Blossom and her family returned to the wild.

The next time some Biblical scholars decide to produce a new version of the Bible, I think I should be consulted. A verse or so in the Bible mentions the concern of God for all creatures — even the sparrow. Gospel songs have been written about that. I can remember, when I was a young child in Church School, singing:

"I sing because I'm happy,
I sing because I'm free!
His eye is on the sparrow
And I know he watches me!"

I happen to know that God cares about very small skunks as well. No other way can I explain this adventure.

One day I was working with a local television representative who was beginning a documentary about the Sanctuary. She had taped the examination of a badly injured bear cub. She had filmed the work of our veterinarian, Dr. White, giving rabies vaccinations to about fifty raccoons. She wanted a record of some animals being set free. I took a kennel which I thought contained four skunks ready to go free and off we went to the bush.

We drove along the highway to a gravel road turn-off; along that to a dirt road turn-off and along that for about five kilometers, where the trees were thick and the berry patches and wetlands are extensive. And where few cars ever venture. I carried the kennel back into the bush. She was ready with her camera. I opened the wire door and the skunks trotted out to freedom. Not four skunks. Five! I had not realized that the last little skunk was in the kennel — nor do I know to this day exactly how it got there. But it was much too small to survive in the wild. The other skunks stayed around, hunting, ignoring us. The little one scurried away into the bush and I could not find it.

Later in the evening I drove back to the release site. Though I hunted and called, not a single skunk was anywhere to be found. During the night I thought about the little one and felt both stupid and guilty . . . I should have checked more carefully. A dozen things I should have done. The fact remained, one very small skunk was out in the bush alone and would not likely survive.

Since the next day was Friday, I knew that I would spend it driving. In the morning I would go to Sobey's grocery store in Parry

Sound to pick up the outdated meat they always donate for our wolves and coyotes. I would come back to the Sanctuary, drop off that meat and then drive down to Bracebridge to take the week's mail to the Sanctuary treasurer. That done, I would drive down to Port Sydney to pick up more outdated meat from the Foodlands store (our wolves eat a lot!), and then on up the highway to Huntsville to buy grain and whole corn and oats for the moose and deer. And that is what, that particular Friday, I did.

Driving toward home I thought that perhaps the two dogs I had with me were getting bored with the long excursion and might appreciate being dropped off at home before I took all of the morning's gatherings to the Sanctuary proper. And, a cup of coffee would do me no harm. So I turned up the side road to my lane, up the long, long lane and put the dogs into the house. Then I decided I would wait for my coffee until after I had done the drop-off at the Sanctuary. That way, I would be able to sit at home and relax . . . .

I went back to the car, backed it around and drove down the lane. At the road I stopped.

A very little skunk was trotting along the edge of the road. Our little skunk!

During my morning's errands, I had traveled much of the entire map of Muskoka. The little skunk had traveled along the dirt road and the gravel road for over five kilometers. We had arrived at exactly the same spot at precisely the same time.

You explain!

I picked it up. It was tired and hungry but safe. I wrapped it in a blanket and took it home. It stayed with me until spring.

Perhaps a revised Gospel song saying: "His eye is on the skunk", would not have universal appeal. However, in any new version of the Bible, I believe the book of Genesis should include the words: "And God created skunks". I think so, anyway.

Benji

3 – All Those City Raccoons

The phone rang. Since the phone rings several times a day, especially in the springtime, I answered automatically: "Aspen Valley Wildlife Sanctuary."

The voice was completely unknown to me. "I understand Ricky is coming to your place isn't he?

"Ricky?"

"Yes — Ricky." The slight impatience in the voice indicated that I ought to have known who Ricky was. I didn't.

"Haven't you seen the newspaper?"

No. Newspapers are not delivered every morning — not away out along a gravel road in the back bush.

Later, I was given a copy of the Toronto Star newspaper. On the front page was a picture of a small raccoon clinging desperately to the side of a building. The story read:

> What pint-size actor did his best to impersonate King Kong on a downtown building yesterday while hundreds gasped in amazement?
>
> Why, Ricky the raccoon scaling the Royal Bank Plaza!
>
> The black-eyed bandit held up traffic and turned heads while a Toronto Humane Society rescue crew tried for two hours to rescue the varmit from the 23rd floor of the 40 storey building at Bay and Front streets.
>
> The three month old raccoon began scaling the building at 11 a.m. "When we arrived at 11:30 a.m. he had reached the 16th floor," said Humane Society general manager Michael O'Sullivan. By noon about 200 people had gathered in the rain at the foot of the building to watch Ricky scramble up the east side.
>
> Plaza staff produced a tarpaulin and some 30 people hoisted it above their heads, ready for any strong winds which might blow the

beast from the building. The Royal York Hotel brought out cushions and foam bedding to provide for a soft landing, just in case, and cups of coffee appeared to help rain-sodden volunteers keep up the rescue effort.

At 1:30 p.m., O'Sullivan and his 3 man crew climbed out on the roof and lowered window-washing equipment scaffolding down to the raccoon, who by this time had been dubbed:"Ricky" by the crowd. He climbed aboard and was hoisted up to the roof where the rescue crew popped him into a box and whisked him off to the Humane Society on River Street . . . .

Ricky was safe and in good condition, will stay at the Humane Society for several days before being taken north of Barrie and released into the wild.

Well — that explained the phone call. Several days later, Ricky did arrive at the Sanctuary. He was obviously older than three months, he had no love for human beings and, I was positive, any raccoon who could survive in downtown Toronto, would have nothing but sheer joy at finding himself in a quiet wetland, full of fish and frogs and berries and all those edibles which make a raccoon healthy and happy. I took him in a cage away back into a swamp, set the cage down and opened the door. And I saw the happiest raccoon in the world trot away to the life a raccoon is meant to live.

According to some statistics, if the population of raccoons in Toronto and raccoons in Muskoka were compared, Toronto would win by a million or so. If we were to compare the amount of environmental damage done by raccoons in Toronto, or humans in Muskoka, Muskoka would win — no competition. Toronto, however, has not experienced a raccoon invasion. Once, the land by the lake, the ravines, the rivers, the woods and the ponds was the natural rightful home of raccoons; hollow trees for nesting, streams and meadows for food, quiet space, where the season passed naturally over a peaceful place. Then humans arrived — the trees came down, the streams were tunneled underground, the meadows were covered in cement — cars — noise — garbage. Chaos was introduced at such a tremendous speed that raccoons could only do one thing at which they excel: they adapted. The dark safety of hollow trees gone, they simply moved into attics and chimneys. No food? Garbage! Tons and tons of garbage!

Many humans regarded the raccoons as, at most, dangerous intruders — at least, a bloody nuisance.

I often wonder if the raccoons are basically content in the roar of the city; certainly they do thrive. But could it be that, deep in every raccoon, is a voice which insists: "You are meant to live in green, wild places, where trees grow tall beside streams and swamps, where the only sounds you hear are the wind, the storm and the voices of owls and foxes, of deer and wolves."

Of course, I do not know. I do know that time and again I have been given little raccoons, born in the throbbing life of the city, and have been able to watch them go free into the green depths of the wild. Joyous! Content.

Benji was born in a small ravine somewhere near Bloor Street in Toronto. The ravine bordered a playground often full of children. We do not know what happened to his mother, whether she was killed or simply long delayed when she was out hunting, but Benji, small, alone, wandered into the playground. A group of boys saw him, dropped their baseball, swarmed around him and picked him up. A pet! A pet! One boy took him home. The parents were agreeable — at first. They put him into a small cage. They fed him.

How long during that night, did Benji's mother search for him . . . calling — and calling?

Jeremy, too, was born in a wooded ravine. Again, we do not know how, or even if, he was orphaned — though it does seem likely. Small, hungry and very sick, Jeremy crawled up from the woods, across a lawn and into some shrubbery where he collapsed and lay very still. All night the rain poured down. In the morning, a pretty young lady found him and cared. She picked him up, wrapped him warmly, and took him home — seven stories up in an apartment building.

Little John was born in an attic or chimney. Considered a nuisance, unwanted, he was simply taken to a garbage can and tossed in. The garbage man found him in the can, and at least cared enough to send him to the Humane Society.

Soon the parents of the boy who was keeping Benji in the cage, grew tired of him. They knew he would grow up and not be the sweet little pet anymore. They phoned our Sanctuary.

"Would you like a little raccoon?"

"Yes." We made arrangements to meet at the parking place at the junction of Highways 400 and 7. The boy's father found it rather incomprehensible that anyone would care enough about one little raccoon to drive so far south to pick it up, but the boy himself, I rather think, understood better. Earnestly he explained to me, "He is very clever. And he likes ice cream".

I told him that I was glad Benji was clever because he would have to learn to hunt bugs and berries, frogs and fish, clams and all sorts of little squigglies from the bottom of the pond. Then the boy said "goodbye" to his raccoon and hopped into his dad's car. Tired, Benji curled up on my knee and slept all the long, long way up the highway to Muskoka.

Jeremy's trip to the Sanctuary was more exciting. His pretty girl decided that an apartment was not the best place to raise a raccoon. He roamed through every room, breaking dishes, upsetting plants, making puddles and piles behind chairs and under the beds. She put a call through to the Sanctuary: "Would you like a baby raccoon?"

"Yes." He was put into a small, dark kennel — had a trip on the thundering subway to the car of a friend — and the long trip up the highway — to Muskoka.

Little John, in a cage at the Humane Society, simply waited. At that time, the Society did not have a wildlife programme and Little John was kept where he could hear the yelping dogs, ate little and slept little and the people wondered what they were going to do with him. A chance visitor, who knew of the Sanctuary, saw him and (without the usual phone call), put him in a big cardboard box and brought him to us. When he arrived, the answer was, of course: "Yes!"

When the cardboard box, set in the middle of the living room floor, was opened, Little John put his hands on the edge of the box and looked out. Benji and Jeremy sat on their fat little haunches and looked at him. For a few long, long moments they looked at each other. Only they understood what their thoughts really were — but, we do know, that they decided to become friends.

Benji was the biggest coon — his coat was beige-grey, his mask as black as ink and the hair around his nose creamy white. Jeremy was easily the most beautiful — all silvery with a dusting of black over the tips of his fur, his mask very black and the white marks very white. However, Little John was a raccoon who looked as though he had been eating nothing but garbage. His coat was dry and faded. The rings on his tail were so faint they were barely there. His mask was a dingy brown. His whiskers were broken, his legs were bowed, his eyes were dull. Perhaps that voice, telling him he belonged in the green and quiet places, was almost too faint to be heard.

Benji, Jeremy and Little John came to the Sanctuary in the 1980's — that seems so long ago now!

Away back then, I was here alone and beginning to learn about raccoons. I made hundreds and hundreds of mistakes. Certainly the raccoons had much too much contact with humans.

I allowed them in my house. Sometimes they even ate human food. Though these experiences are now strictly forbidden to raccoons — especially by authorities who make all the rules — the raccoons survived well, grew up and lived long and full lives out in the bush. However, I do want it to be known, that though I was not necessarily acting wisely, the raccoons were busy teaching me all sorts of things about raccoons. And what they so thoroughly taught me has been used to the advantage of hundreds of raccoons since then. So, while I am sharing with you the adventures I had with these three, please do notice — I was not teaching them very much. They were teaching me a great deal!

Little John was so small and so homely and looked so utterly abandoned, I picked him up and hugged him and even kissed him — a process which animals do not necessarily like. Still, he cried in his throat and thrust his nose against me and snuggled tightly into my arms. He had been a long, long time with no living contact. And, I still believe, he felt loved.

At first the little raccoons were given Pablum and goat's milk (I did know that cow's milk can kill baby wildlife). After awhile, they ate dog kibble and whatever fruit I could persuade them to try — they liked: strawberries, raspberries, apples. Still, while feeding from a bowl was all very well, they had to prepare for the day when they would live in the wild and have to find their own food.

Every day we went for a walk to the little stream which wound its way through the valley. (Since then, the beavers have dammed it up and it has become a small lake.) They seemed to know what to do: they felt around the stream bed with their hands and found bits of roots and weeds and bugs, and drew them out and ate them. They saw the startled frogs hopping away, and chased them. And learned to catch them. (One could not help feeling sorry for the frogs; caught, they would be rolled between the hands of the raccoons until they were as long as wieners, but still alive — and then eaten long — while still alive. They found clams and knew how to open them and eat them. When we took them walking in the fields, they found strawberries in season and raspberries. They dug around in rotting stumps.

They climbed trees — all of which they did without instruction from me. I know very little about hunting for food.

Preparation for survival during the winter months, when white blizzards will whirl out of the low, grey skies and bury the wilderness deep in snow, must be a driving force in the lives of all wild creatures. These little raccoons knew nothing about temperatures that would drop so far below zero that streams and ponds wold be frozen hard, and bare trees would crack like guns in the cold and yet, some indicator in their bodies told them that they must begin to eat and eat and eat so that they would grow big and be covered in layers of fat so that they would survive those long, lean months. So they ate and they ate. They hunted earnestly. I still fed them — I knew all about winter.

Still they were young raccoons and they played. Out in front of the house an apple tree had stood for a hundred years. Though it bore few apples, its branches were old and thick and spread across the split rail fence and over the tall grass to the meadow.. Using it as a gymnasium, the coons would scramble up its short, fat trunk, bounce out the sprawling limbs, sometimes on top, sometimes hanging upside down, tumbling, sometimes thudding to the soft ground. Undaunted, they would climb once more to the tip-top branches where, clinging, they would survey the valley and the fringing woods, and sway in the wind.

They played in the barn. The barn is huge, its vast roof crossed with gigantic beams, its tall sides crisscrossed with supports and wooden ladders reaching right up into the dim peaks. In the summer, the barn was filled with sweet smelling hay. The coons climbed, they tunneled, they wrestled, they played hide-and-seek, they hunted mice, they pounced, they were finding life very, very good.

Thirty feet above the floor of the barn hangs an old log, suspended years ago as a lifting device. Jeremy and Little John discovered it. One at each end, they would point their noses toward each other and charge, their claws scratching and scraping along the bark. Gathering momentum, they would meet in the middle and one —

always a different one — would have to swing under and hang upside down to keep from falling. This exciting game was repeated over and over again. Benji's game depended more on strength than on agility. He would take his place on the top bale of a pyramid of straw and defend it against the repeated charges of the others — down they would tumble, recover their feet and scramble up again. He always won.

Shadow River, a waterway which is too large to classify as a stream, but really not large enough to be a river, flows into the lake just at the edge of Rosseau Village. One claim to fame the river does have: Pauline Johnson, most distinguished of Canadian poets, wrote one of her most well known poems about the wonderful reflections of the great trees, exceptionally clear because the water in Shadow River moves so slowly and deeply between the curving wooded banks. However, if Shadow River is followed deeper into the bush, the banks close in and the tall trees are held back by the huge rocks, forcing the water into a narrower channel. A fair distance from the lake, when the river is still a stream and the rocks begin to close in, the water tumbles over a series of waterfalls and rapids — none of them tall or swift or dangerous — simply excellent places for little raccoons to learn about fishing. So, one afternoon, I took the three raccoons to a place in the bush where Shadow River slides over a smooth ledge, rippling over the mossy rocks to fall, with a rim of froth and foam, into a dark pool, shaded by tall evergreens.

Never before had the raccoons seen such fast moving water. Wading in, Jeremy stood with his legs braced against the current — and then he slipped, slithered, struggled a moment — let himself go, sliding down the rock, over the ledge and into the pool. Under the water he went and then up to the surface like a little boat, he circled, swam and finally made for the shore. Soggy wet, he pulled himself up on the rock, shook — and decided he did like swimming. He was back in.

Watching, Little John was more cautious. He climbed a tree and watched. And watched. Only when he saw that Jeremy had discovered what raccoons were supposed to appreciate about water, did he come down. Jeremy was poking under small stones, turning them over, lifting them, and obviously finding things to eat. All sticky with pine-gum, Little John followed, began investigating under the stones for himself, found all sorts of wet grubs and greens which were good food, and, perhaps deciding that Jeremy shouldn't have all the adventures, he finally launched himself into the water and swam.

Benji took the entire adventure more seriously; he did not let the current push him over the ledge. He did not waste time swimming. The water and the moss were full of food. Benji hunted, and he ate.

Let other raccoons play games that were merely fun. Benji put thought into all that he did. There was a time when, growing fat toward winter, he did not want to undertake a long hike with people, the dogs and Jeremy and Little John. He followed a little way up the path, apparently decided that too much energy was required, so he sat himself down under the shade of an evergreen and watched us all walk away. When, an hour later, hot and tired, we returned down the path, he was waiting beneath the tree. Fresh and cool, he scampered away with us the short way home.

He thought about his endeavours. In the house was a green, wooden rocking chair. If a person, wearing slacks or trousers, sat in that particular chair, the pockets would be very accessible to the deft hands of a thieving raccoon. I began to find coins hidden behind books, piled in corners, under the sofa, and I began to wonder and finally caught Benji in the very act. As quietly as a thief who made his livelihood doing it, he could slide his hands into any vulnerable pocket, take out the shiny coins and hide them.

His other amusement was flushing the toilet. He would sit on the seat, flush . . . watch . . . wait until all was quiet — and flush again. The toilet seat had to be not merely kept closed but weighted heavily enough that Benji would be discouraged. In the house, Jeremy and Little John behaved like ordinary raccoons — exploring cupboards and closets, opening refrigerator doors and stealing whatever food they wanted, toppling plants, dragging books out of bookshelves — everything which makes having raccoons in the house the height of folly. We built them a large outdoor hutch from which they could come and go as they pleased and learn to be real raccoons.

Winter came with a sudden storm. The raccoons sound asleep in the straw all through that night seemed to be unaware of the outside tumult which was turning their world white. When morning came, they poked their noses out — tentatively — and wondered. They had never seen snow before. Benji turned back immediately and dug himself under the straw. Hesitating only a moment, Little John followed him. Jeremy ventured out to the roof of the hutch, held up one paw — then another — discovered that the snow was cold but that he could not keep all four paws out of it at once — and then back, with the others, to the warmth of the straw.

Only later in the day, when the sun began to shine dimly through the clouds, did the raccoons allow their curiosity to lead them out into this new world. Snow? Perhaps as long as the wind was not howling, snow wasn't so bad. They made a long trail up to the barn. And soon the trail led across the deep snow of the meadow to where the pines stood tall along the hillside. The coons spent more and more time in the pines, perhaps finding hollow places in the big branches where they curled up and slept. When, some months later, warm winds began to blow over the woods and the snow was melting away, the raccoons began to move about.

Little John, the raccoon who had begun his life feeding in the garbage cans of Toronto, wanted the woods, alive with the sounds of wild things and wind and rain. He hunted for himself. Though we saw him occasionally, enough to know that he was doing well, he became a wild raccoon.

For Benji, the conflict between the house and the woods was much greater. He came home. He played with the dogs. He slept with them, a warm ball of grey-brown fur in the midst of black Labrador retrievers. Sometimes he slept in the bed with me. For several days he would be a house coon. Then the green woods would lure him away and he would vanish for a little while.

Just when I began to believe that he had at last become a truly wild raccoon, I would hear a rattle at the door, and Benji would be back again.

However, he did leave. The last night he slept on my pillow between me and a big Labrador. He slept heavily until about 2 a.m. when he ran his fingers through my hair and across my face, then, back to sleep. Around 6, he awoke, got off the bed and began to uproot the geraniums.

Obviously, the time had come for all of us to abandon any effort to sleep. The sun was rising, bright and warm. I made a cup of coffee and sat outside the back door, drinking it. The dogs followed their noses over the delicious night-time trails. Benji peered out the door, sniffed and trundled down the hillside to the stream. Into the shadows he went, hunting, groping at the stream bottom, almost pretending to be serious. He kept glancing up at me, like a child assuring itself that the adults are paying attention. He frisked in silly circles — ran up a tree and down again — resumed his hunting.

Then, suddenly, almost as though he heard the call of the green woods, he turned and trotted away. He didn't look back. Though we saw him again and again, and knew he was surviving well, he never did return to the house. Nor were we permitted to touch him. He was wild.

Jeremy stayed on living in the hutch. Though he roamed free, we were not certain he was hunting on his own and so we did leave kibble for him. With spring, of course, come more baby raccoons. As we had given care to Benji, Little John and Jeremy, so we cared for the new little ones who were given to us. These raccoons had one real

advantage which we had not been able to give to the first three —
they had Jeremy. Jeremy was a male. (The biology books will tell you
that, even in a natural situation in the wild, male raccoons do nothing to help raise their young. We began to learn then that the fact that
animals cannot read is a distinct advantage). When the little raccoons
were old enough we could have them outside, Jeremy took over.

We committed the dreadful anthropomorphism of dubbing him
"Uncle Jeremy". He adopted the little coons, taking them everywhere
with him — to the woods, the stream, through the fields. He helped
them catch clams and frogs and minnows. Every evening, he led them
home again and left them at the house. While we took them in for
more food and a warm bed, Jeremy would vanish away, back into the
woods to do whatever wild raccoons do during the night. When that
litter of raccoons no longer needed him, Jeremy himself became a
wild raccoon.

I wonder if any of them ever remembered the city.

I do not know exactly where in Toronto Moon was born, or how
a very, very small raccoon came to be the mascot of a singer and her
band. Because she had a dog called Star and because she enjoyed
words which rhymed, the raccoon kit was named Moon. Moon, the
Raccoon. Though I never did hear it, I understand that the raccoon
had earned a song of that name, a song which was being sung in
assorted clubs and colleges across Ontario. I do wonder how the
clamour of the music and the clubs, added to the noise of the city, was
appreciated by one very small raccoon who should have been hearing
only the sounds of the wilderness.

By September, he was causing too much disturbance in a Toronto
apartment, too much damage and too much noise for any musician
who wanted to compose music. And so, one day in late September,
Moon came to live at Aspen Valley. He came in a very small cage in a
car. His singer set the cage on top of the woodpile in the lane.

"I want to take come pictures of him before I go." This pictures-at-the-last-moment syndrome is very common but not one I really understand; is it not more realistic to take the pictures at home where the raccoon has been living? Anyway, she opened the cage door. Before the camera was focused, Moon was out of the cage and in one frantic dive, deep down amongst the crevices of the log pile. Deep, deep in the log pile. He would not come out. Coaxing, bribery with food — no! He stayed in the pile. She had to leave with no pictures.

Knowing what had frightened Moon was not difficult to determine. He was accustomed to cars and trains and subways — but not the wind crossing the meadows, or the murmur of trees. He knew one dog — Star — now he heard five dogs, barking. If we had dismantled the log pile, he could have bounded away before we could have caught him — easily. And, catching a frightened, unwilling raccoon can be a bloody mess. The best thing was to leave him and let him come out when he wanted to. He did. Some time in the middle of the night.

My dogs are so accustomed to wild animals that they either totally ignore them or else agree to become friends. However, very good friends of mine had asked that, while they took a vacation to Europe, I would look after their dog. I like dogs. I liked their dog, a huge Airedale. Ianto was the fifth barking dog. He did not know he was supposed to tolerate, if not be friends with, raccoons.

During the next few weeks I saw Moon only three times. Once, drawn by the fact that Ianto was barking frantically at the foot of one of the evergreens which line our roadway, I went to drag the dog back from the road. I saw Moon on a branch high up in the trees, ears flat back across his skull and hissing. Once again, several days later, I heard Ianto's frenzied bark, this time down by the pond. Ianto, ordinarily afraid of water, had waded out to his belly. Just beyond the tip of his out-thrust nose, half in the water and half clinging to the trunk of a witch hazel bush, was Moon, furious and letting the dog know it. Once more, Moon vanished into the wilderness. Several days passed before I heard, for the third time, the frenzied barking. Ianto was at an old, huge pile of rocks across the other side of the pond. I ran across the bridge. Moon was deep in a rocky cave but Ianto was a large dog and he was scrambling at the rocks, able to move them. I dragged him back to the house, shut him in a kennel, and told him: "That's where you stay."

And I meant it!

To Kate, one of my dogs, I said: "I'm going to take some kibble over to Moon, just in case he isn't hunting well." However, as I was walking, bowl in hand, across the bridge, I met a very self confident Moon coming down the hill toward me. I turned. He followed me up the hill, passed the kennel, onto the porch and in the door. He ate. He curled up on the couch and went to sleep. What did he know? That the one dangerous dog could no longer hurt him? That the other four dogs were perfectly friendly? And how did he know?

We know little about animals, even about animals as common as raccoons. Moon had endured a very traumatic few weeks; separated from the person who had raised him, the long trip from Toronto, the unfamiliar valley, the hostile dog now, he slept. He began to run a fever. He would not eat. I forced water down his throat. For a day I worried. By the second day I was afraid he would not live. He slept, hot and unmoving. After almost a week of sickness — both the veterinarian and I had tried everything we knew or even imagined — Moon got up — wobbled to the door and asked to go out. My first thought was simply: "No way!" He persisted. Finally I thought: "Okay, we'll see what he wants. But I'll stay with him."

Weakly, he tottered down to the pond. He was so weak I was going to pick him up and take him back to the house — but up there, suddenly, a great commotion erupted. I heard it. Made a dash up the hill — I could find nothing. When I returned to the pond, Moon was gone. Though I searched, I could not find him.

For three days I wondered about and worried about the sick, city raccoon, lost and frightened in the bush. On the fourth day came a persistent scratch at the door. I opened it to a thin, but bright-eyed raccoon who came in and made himself at home.

What antidote had he known and found in the wild?

Ianto's owners returned and were glad to take him home. He was glad to go! And so I had only my own dogs, several cats and a skunk named Dugan living in the house. Dugan had been another victim of humans who want a wild animal as a pet. Unlike raccoons, who can be reintroduced to the wild, the skunk was handicapped because his only means of defense, his scent glands, had been removed. He could no longer spray. He was completely vulnerable to whoever happened to want to eat him. And so, Dugan lived in the house, ignored by cats and dogs, sleeping mostly in the cupboard under the bathroom sink, using a littler box, as little bother as any creature can be. His life was, I suspect, somewhat dull until Moon moved in. They ate together. They slept together. They wrestled. They played. When Moon went outside, he went along.

I have walked in the bush with bears and with deer, watching, sharing, learning. Thus, one day, I decided to record exactly what a day in the bush with a raccoon might be like. Accordingly, I took my camera and, instead of having Moon follow me, I followed Moon. He led me across the back field and through the woods, up the hillside. The walk was slow — he went many places I could not follow. Mostly up trees. He spent half an hour high up one tree, filling his mouth with wild cherries. At that time of year, maple trees were interesting to him only as challenges to climb up fast and down again. On the forest floor, he wandered casually, nose down, pausing to grubs at roots, to turn over stones, to scratch at the lichen. He spent much longer at the pond — a clam or so, and an unfortunate frog. (I do wish raccoons were not so fond of frogs!) He wandered in a rather circular direction and we ended back at the house. There, he curled up on the sofa with Dugan, and they slept together in a warm, furry heap.

I did get some excellent pictures.

His household routine was much more original. He searched pockets, he opened drawers and spread their contents all over the room, he uprooted plants. He would steal a plug from the bathtub so that, before one could bathe, one had to hunt all over the house to find the plug. He was relatively good at keeping off counters and tables; he could not resist waste paper baskets.

Madge, a most excellent friend, came to stay with me overnight. At that time, my little house actually had two bedrooms: mine, and a spare room where the occasional overnight guest could stay. (One had to be rather brave to stay at my place.)

Madge left her watch and her charm bracelet on her dresser. Still, after what I had thought was a relatively uneventful night, the charm bracelet disappeared. We accused Moon. He was completely unconcerned. He even looked innocent. We searched and we searched every bookcase, the laundry basket, the woodpile, the cupboards and closets — everywhere. No bracelet. I resorted to begging Moon to confess.

Several months later, a nephew who had been staying in that second bedroom both before and after Madge's visit, came into the living room dangling the charm bracelet on the end of his finger.

"Can anyone explain to me," he demanded, "how this came to be in the second drawer of the dresser, under my underwear?"

Moon understood a number of words but obeyed only when he wanted to. One morning, in my armchair by the window, short-tempered for a number of reasons I cannot now recall, I was enjoying a second cup of coffee. Moon wandered restlessly into the room. He decided to climb up on the television set.

I said: "No, Moon! Down!"

He got down. He trundled over to the table and decided that perhaps there was good food away up there. He started to climb up.

I said: "No, Moon! Get down!"

He did. He crossed the room to my sewing basket. He inserted his fingers and raised the lid.

I said: "No, Moon! Stop it!"

He sat down on his fat behind. He eyed me. He flipped the lid again.

"Moon!" I said, emphatically, "No!"

He flipped the lid again — and again — and again — solemnly watching me. Until I laughed. Then he left it alone.

March was one long month of blowing winds and blizzards. Even the dogs went outside only when nature demanded and they never stayed out very long. I kept the fire in the woodstove going. Moon and Dugan slept as near to it as they could get, on a woollen blanket, in the wooden box. I, of course, had to go in and out to tend the creatures in the barn, to bring in wood, to go in to the store for food. After one such trip, when the snow was whirling around the house, and the wind was howling in the darkness, I closed the door and said to all the creatures:

"That's it for tonight!" and put another log on the fire.

But Moon would not settle. He paced. Front door. Back door. Front door. Back door. I said to him: "No one in its right mind wants to go out tonight!" He scratched at the front door. He scratched at the back door. He would not settle. The cats slept. The dogs slept. Moon paced.

Back and forth.

Finally, exasperated, I said: "I'll show you!" I picked him up and carried him to the back door and opened it. The blizzard roared in. But there, crouched against the door was one little skunk, ice-covered and terrified. I do not know how he managed to get out. I let Moon go, picked up Dugan and cuddled him by the fire until he was warm again. Moon waited. Then the two of them climbed into their box and went to sleep.

Moon cared about his friend. His friend, the skunk.

When spring finally came, Moon grew very, very restless. Sometimes he would join the other raccoons in the barn; sometimes he simply wandered off by himself. He would be gone for several hours. Then for several days. When he did come home, he was never particularly hungry, so, though I missed him, I did not worry about him. All summer and fall, he was nearby. When winter returned, he was back in the house — until spring.

After the third winter, he simply disappeared into the bush and did not return.

City raccoons? Yes. But they certainly adapted very well to country living!

4 – Moses and the Sixth Commandment

The conversation would be pure fantasy and highly anthropomorphic, but if I could question Moses in whatever "heaven" he is now inhabiting and ask how he had enjoyed his life, I am convinced he would answer, "It was great!", and add, because he was a raccoon, "And you enjoyed it, too." "Right, Moses, I did. But it was a near thing. If you happen to be able to confer with Doris up there, you might demonstrate enough humility to thank her. Your good long life was primarily her decision. To which, mind you, I agreed."

The fantasy being now complete, before you meet Moses, you must meet Doris.

Moses

Away back at the beginning of the nineteen seventies, this was a quiet, green valley. A small bay horse and a pinto pony grazed in the pastures. Two goats wandered about, eating every new tree I planted to replace those that had been destroyed when my log house burned.

Three black, mostly Labrador, dogs wandered about. One large cage near the barn was used for either raccoons or skunks, as required.

One warm spring day I was sitting in the opening of the big, sliding doors which open into the top floor of the barn, and enjoying the valley, the animals and the quiet — and then, (I must have known life was about to change), I heard the sputter of a motor and saw a peacock green Volkswagon Beetle crest the hill of the dirt road, spin down, be lost for a moment behind the big evergreens and then clamber up my laneway. It came to a stop at the bottom of the slight hill which leads up to the barn doors.

Now what? I could see that the car was loaded with bags of kibble — back seat to the roof — in the trunk — under the nose of the car — tied to the roof. Then the door on the driver's side opened. Doris stepped into my life.

"Hello", she said, I'm Doris Hoare. I've heard about you. I heard you look after raccoons."

I'm still not certain how, away down in Toronto, she had heard about one small raccoon cage in Muskoka, and a large, remote valley. The Aspen Valley Wildlife Sanctuary had not really been thought of, let alone established. The interest in rehabilitation work was still in the more distant future. Except to Doris. No rehabilitation centres existed in the big city of Toronto. Even the Toronto Humane Society had nothing. When raccoons, or any wildlife, was given in at the front door of the Society, it was instantly taken up for euthanasia. Doris did not use the "polite" word.

She said, "They kill them."

However, because Doris had a few caring friends working at the Society, she was enabled to begin her work of rescue. A raccoon would arrive at the front door of the Society — one of her friends would manage to take it and phone Doris who, as fast as her green Beetle could weave its way through the traffic, would arrive at the back door and take the animal. When occasionally she was unable to get to the back door, her friends would hide the creatures in brown paper bags, and, via the TTC [Toronto Transit Commission], (also illegal), arrive at the front door of Doris' down town house.

She did have a few cages in her garage — the Beetle sat in the driveway — and a few in her basement. She knew she needed a country place for them. That is where I entered the picture.

And that was why the little green Beetle came scuttling over the hill, down the hill and up my driveway. We talked a long while that day — and I began to plan for more raccoon enclosures. Before she left, we unloaded all the bags of dog kibble and stored them safely in the lower part of the barn.

Over the years that followed, Doris brought many raccoons — small ones with their eyes still closed, unweaned kits, some that simply needed a place to live until they were old enough to be set free, some that could go immediately.

And then, Moses.

That adventure began with the usual phone call. Both Doris and I were facing a new situation.

"I've just been asked to take an adult raccoon," she said, "He's totally blind."

Handicapped. Blind. The opinion of the veterinarian in Toronto was that, since he could never go free, he should be euthanized. Killed. After all, what kind of life could a blind raccoon have being always kept in a cage? Doris and I talked it over. Of course, the veterinarian knew best. He had pronounced his verdict. Case closed. We hung up the receivers. On a raccoon, now alive, we had passed the death sentence. I thought about — not all the fine, healthy raccoons in Toronto and Muskoka — but about one single, handicapped raccoon. A living being.

Down in the city, Doris was thinking, too.

I picked up the phone. Dialed her number. Line busy.

Doris picked up her phone. Dialed my number. Line busy.

Finally, we did connect. And we had changed our minds. Moses would live. And he would show us that handicapped animals can have very excellent lives.

Within a day or so, Doris called again. "I've got the blind raccoon here with me. But I've got more raccoons coming in, so I cannot be away long. You'll have to meet me part-way." And, since I had confessed to having received the first beaver kit I had ever had, she added, "and bring the beaver. I want to see him."

"Yes, Doris. And you'll just love him." My enthusiasm was greeted with a slight silence but then we made plans to meet. In the town of Belfontain was a small brick church, on the main highway, beside a cemetery where trees gave shade and flat gravestones were available to sit on. We would meet for lunch. She would make the lunch and bring it along. All set.

It was some time later that I was to learn that the small silence which had greeted my enthusiasm for the beaver was really apprehension: What if I came to like beavers better than raccoons? However, the future holding what it has, perhaps she sensed where my addiction would eventually lie. But I still like raccoons very much, indeed. At the time, I fixed up a small carrying kennel for Swampy, the beaver kit, took the bottle of formula, and set off for the flat gravestones. Because beaver kits are so attached to their parents, and hence to the caregiver, taking him with me would be far less traumatic than leaving him alone for many hours.

I found Belfontain, the church, the cemetery and Doris sitting on a gravestone, waiting. She met the beaver kit and acknowledged that he was wonderful. I met the blind raccoon and was glad we had not allowed him to be killed. We ate our lunch, talked for a while and then Doris announced: "By the way — I brought you a skunk, too. I didn't think you would mind."

"A skunk? Descented?"

" I really don't know. He was found walking through the bus terminal on Bay Street in Toronto. He hasn't sprayed anybody yet." That was all we knew about the skunk. I did take him home. He did not seem particularly tame and, finally, he went free.

However, back at the cemetery, we arranged my car with the beaver, the skunk and the blind raccoon, and I started home. As I drove northward I saw dark clouds ahead — the sunlight faded — and, the snow began.

I have always enjoyed speculating about situations in which I find myself: are they absolutely unique? Has anyone else ever watched a woodchuck trying to housebreak a raccoon kit? Or a dog and a skunk and a raccoon curled up together on the chesterfield, sound asleep/? In this case, had anyone ever driven through a blinding blizzard with a beaver, a blind raccoon and an "undeclared" skunk in the same car? When I encountered a white-out, I pulled over to the side of the road and waited.

That was how the blind raccoon came to the Sanctuary. From some rather hazy recollection that Moses of the Old Testament was unable to enter the Promised Land, linked with the fact that this raccoon could never wander free in the valley — well, the connection wasn't exactly logical, but the name was good. Moses.

As Moses, he pioneered the path for the many handicapped animals who were to come after him.

The lower part of the barn, which in reality might be classed as partly a basement, though it does have windows looking out over the slopes of rise, was still in its original state: stalls for the horses, a large box stall for the goats, a smaller box stall which had been given over to some young skunks — and no real place for a raccoon. Moses would have the run of the place.

Sight is not the most important of the raccoon senses. Nose down, Moses began to explore. Everywhere. He avoided the horses. While the goats stood aside in some perplexity, he examined their stall. He walked around the perimeter of the wall of the skunk pen. He discovered the wooden stairs which twisted up to the floor above, and went up and down them again. I put out a bowl of kibble and a bowl of water. He had no trouble finding them.

That night, when I phoned Doris, I told her that I suspect our decision had been correct: although I doubted that Moses could survive in the wild, he would certainly be able to live a good life. And so it proved.

Moses, having the run of the lower part of the barn, came to know every inch of it. He recognized all the animals who lived there, avoiding the horses but otherwise unconcerned or faintly friendly. He shared his food with the cats. When I was there, cleaning or feeding or merely enjoying being with the animals, he stayed nearby. I was able to watch him. His nose was always down, tracing the scents. His hands were active, feeling his way so quickly that his progress seemed normal.

His reactions were often very normal, too. One afternoon, a woman came to visit. She was a rather spoiled woman, unused to working very much, so, as she sat on the stairs watching me clean, she talked and she talked and she talked. Her high voice was rather like the rapid firing of a machine gun. I was not watching our wandering Moses. One moment he was exploring a manger, and then, as though he could stand no more, he raised his head, charged across the floor and bit her. Hard! And walked away. She never returned to the Sanctuary. Moses, of course, stayed for years and years.

Keeping a handicapped animal is often supposed to be cruel. Shortly after Moses arrived I was speaking to a group of adults at

Grundy Provincial Park. The day was warm. We were seated under a shady enclosure and I was seated on a convenient picnic table, explaining to them the rehabilitation work in which the Sanctuary was becoming so deeply involved. I talked about Moses, explained that he was blind, and our final decision to keep him anyway. And, I was interrupted by a round of fervent applause from the front two rows. Startled, I repeated, "We've decided to keep him and give him the very best life we can."

Cheers. Finally I had the good sense to look at those two rows of people more closely. They were blind. Every one of them. They understood Moses more completely than either Doris or I ever could. And they most fervently approved of letting him live. Afterward I stayed and talked to the group of handicapped people. They said exactly what we are hearing from handicapped people today: life can be very good.

While rehabilitation work was struggling toward birth down in the city, it was completely unknown in Muskoka. Raccoons and the occasional skunk had been kept as pets; anecdotal stories abounded. "Oh yes, when we were little we had a pet raccoon" . . . or, "my dad did", or "my aunt . . . " And the story about a sweet creature who cuddled in your arms, followed you everywhere and slept on the end of the bed. As seen through the rosy mists of time. But why would one want to release a raccoon? Or keep a blind one? Silly old maid. Still, I had to hire someone to help me build a large, outdoor enclosure for Moses. I hired a couple of men.

I tried not to see that they thought the job was a little foolish. A pen twenty-five feet long, six feet wide and eight feet high, with a shed at the end for shelter. I could see they thought that was all a bit much for a single raccoon. Nevertheless, the enclosure was built. The shed had a wide shelf part-way up the back, a large window above it (For a blind raccoon? Yes. He would feel the warmth of the sunshine coming through), a wide ramp from the floor to the shelf, and lots of warm, dry straw. In the outside part of the enclosure I put a huge, hollow stump that had, obviously many years ago, been torn from the earth because the roots reached out like great crooked arms in every direction. A large hollow log lay on the ground; a tree trunk, slightly leaning, led up to the roof and to another branch which crossed to the opposite fence. He had a round wading pond (straight from Canadian Tire), kept full of water and which had to be cleaned out at least once a day.

Moses loved his pond. With deep and quite sincere apologies to them, we gave him live crawfish and minnows, knowing full well that they had a right to life, too. How nice if life were only logical! He was a good hunter, feeling with his hands, moving swiftly when he found his victim, and crunching it merrily. Sometimes, with a hose, we sprayed water over him. Almost literally, he danced in it, trying to catch the spray, twisting this way and that in his pond, disappointed, and quite vocal about it, when the play had to stop. He climbed his tree to the log where, legs hanging out on both sides of it, he would doze in the sunshine. Often he slept on his shelf in the warmth of the sunshine through the window. On colder days, and in the winter, he burrowed into the straw and made himself a warm den. He dug burrows under the roots of the stump. When we gave him toys, which we did, he would play with them briefly and then turn back to the activities which are more natural to a raccoon — digging and swimming, climbing and sleeping. And eating, of course. Though he liked a few people, he did not want to be handled. His dignity was too great and his own, and we had no right to damage it. Moses lived for twelve years — a very happy and contented raccoon.

Because Moses demonstrated to us that a handicapped creature can live well, we have been glad to give the same opportunity to other creatures, with other handicaps. Perhaps owing to the use of pesticides, blindness does seem to be the most common handicap. We have had other blind raccoons, squirrels and skunks — even a deer.

Ravens, crows, hawks and owls — wings shattered by cars or gunshot — live in huge, natural enclosures. Nor do we need the excuse that these can be used for education, for breeding to perpetuate an endangered species, for fostering other orphans of their kind — though, if need be, these things can be done. We keep them because life is good, and they love to be alive.

One example? Once, I was involved with a blind dog — though no one knew at the time that he was blind. Bandi was a Siberian Husky. I was at the veterinarian's office for something or other, waiting, when the dog was brought in to be euthanized. Killed. He was young, beautiful and his tail was wagging in friendly greeting.

I tried not to interfere, but finally, "Why does he have to be "put down?" (The other euphemism).

The owner shrugged. "Can't train him. Don't know why. Can't use him."

I didn't need another dog, so, when I took Bandi home, I was quite glad that my neighbour wanted him. He already had one dog, a German Shepherd named, Heidi, but he liked Bandi, and so did Heidi. For many years the two dogs were friends, running together through the fields, along the dirt road, through the woods. Bandi was always slightly behind Heidi, his head at her flank. Not until Heidi died of old age, and Bandi was alone, did we sense something was wrong. One day he ran in front of a slowly moving car and was hit and hurt. Not badly. We took him to the veterinarian. That's when we discovered he had been blind since birth. Yes, he had needed Heidi for his good life, but he had had a good life.

Just one more example — or two.

Amos arrived as a very, very tiny skunk, all alone. He had been found near a golf course somewhere down by Pickering. At first, because he was so young that his eyes should have been closed, we were unaware that they would never, never open — at least, not to sight. He grew. Other little skunks his age had wide, shiny black eyes.

His stayed closed. We were still not too anxious. He moved about with the kits from another litter. They adopted him quite nonchalantly, playing, wrestling together, crowding around the Pablum dish. Still, his eyes did not open.

Then one eyelid did open — just a crack. The eye beneath was opaque. Blind. While leaving him still with the other skunks, we did know that when the time for them all to go free would mean that, for him, we would have to make other plans. Meanwhile, he was quite content.

We have a huge beaver enclosure. A creek flows through the middle, between rather steep banks, but most of it has long grass, with hollow logs, an A-frame for shelter (the beavers never use it!), and the branches of trees which, when they still had leaves, formed a green jungle. Now, the grass growing thick and tall through them, they are simply a rather wonderful tangle, where a skunk should be more than content. We gave Amos the freedom of the beaver enclosure; beavers are peaceable folks, and Amos would be able to ignore them.

He did. The beavers swam in the water, burrowed along the banks, grazed on the grass. Amos lived in a hollow log, dug burrows on the surface, hunted grubs in the long grass and lived very well, indeed. If he did come up against the fence which kept him in, he seemed to regard it with complete indifference, and simply changed the direction of his progress. He found the one isthmus across the creek and traveled it. He never fell into the water. During the winter he stayed in a deep burrow which he had dug for himself. There, quite content, he trundled through all the years appointed for a skunk to live, and then, one day, he went to sleep, quietly, and did not wake up. An untroubled life.

I do not know how the following story will end. We had a call from the Urban Rehabilitation Centre in Montreal.

"We have a little fawn — a couple of days old. It has cataracts on both eyes. Blind. They say they can't do anything. Will you take her?" Yes. What other answer is possible with the now deceased Moses and Doris watching from "Wherever"?

The fawn is blind, and, like all deer, easily stressed. With the other fawns, she was frightened.

We have fixed up an enclosure in the barn for her; the darkness she does not know about. She does hear, and like, the movement around her. She eats well. She responds to the sound of a human voice and comes to have her ears rubbed and her nose stroked. Sometime this month, a veterinarian is coming from Toronto to operate on her eyes. This is a procedure which has never been done on an easily-stressed deer before. Worth the risk? We think so. The veterinarian is hopeful. It would be so easy to simply "put her down", as we have been advised. No!

Back to Moses. I suspect he is watching us authoritatively from "heaven" and, having met his namesake, and, no doubt able to recite all of the ten commandments, has written out number six "Thou shalt not kill" and is displaying it firmly. And Doris is standing right behind him.

5 – A Bittersweet Story of Two Beavers

Over the hundreds and hundreds of years, beavers had kept the wilderness alive and green, building ponds which grew into lakes, turning the course of streams and rivers, storing water so that the trees and the plants could put their roots deep into moist soil and drink and grow until their topmost branches almost reached the clouds. To the ponds that the beavers built came all the wild creatures, the animals — bears, moose, deer and wolves, raccoons and skunks — all of them to drink and eat. And the birds came and the fish and the insects. Over all those hundreds and hundreds of years, the beavers worked and the wilderness was thriving. Nature was as it had been created.

But now, so many of the trees had been cut; over the far horizon glowed the lights of a city. Long ribbons of highway lay across the land. Cars flowed by. Near one small stream, where beavers had quietly built a dam, a pond and a lodge, was a human dwelling. In that house were human beings — humans who did not care much about beavers.

Down in the dim warmth of the beaver lodge were three beaver kits, small and round, pressed closely between their father and their mother. Sometimes they slept quietly. Sometimes they talked to each other, soft, gentle mewing, back and forth, content and at peace. In the evenings the parent beavers would leave for a few moments, swim down the tunnel under the water, out to the pond. There, still very quiet, they fed, grazing sometimes beside the pond on the green grass, or else on the lily pads or the cattail roots. They would repair any leakage from their dam. Perhaps they would take a few moments simply to swim together, around and around the pond, diving over each other, playing "push", rolling — until the time came to go back to feed their kits and to be a family together. They had given no attention to the lights from the human dwelling, nor the raucous noise which came from it.

The days passed quickly. The kits grew. Once in a long while they too swam down the tunnel and out into the pond — not often, and only when the day was fading into darkness. They did not know that the lights in the house meant that humans lived there. They did not really know what humans were. Least of all, did they know how dangerous humans might be.

Until late one afternoon, when the parent beavers had gone down the tunnel and out into the pond to feed, the kits heard a horrendous noise. They could not identify human laughter, loud and shrill. They did not understand the explosions, as though the world was splitting in terrifying thunder. Over and over. And the laughter. Splashing! Shooting! And then silence. A long, long silence. The humans had gone away. Their parents did not come home.

They waited. For long, long hours they waited. They grew hungry. All the hours of darkness went by, and, through the little vent in the top of the lodge, a faint light filtered. Anxious, wondering, the kits huddled closer together, mewing. Still they waited. All day and another night. Hungry, now, and growing desperate. Outside, human voices. But quiet voices.

Had he not been so desperately hungry, (and perhaps a little anxious), the little beaver would never have moved. But he was desperate. Very, very quietly he left his siblings, pressed close to each other, watching him, motionless — and he slipped into the water, down the tunnel and very, very slowly, out into the pond. So slowly he made barely a ripple, he moved in the shadows. And he heard those gentle human voices calling him, mewing, almost as his mother had — almost but not quite. He was afraid. He slipped below the surface of the water, back down through the tunnel and back to his siblings. They nudged him, greeting him, wondering if he had found food.

No. They were hungry now, so very hungry.

Again, the next evening, he heard the voices and again he slipped down the tunnel, and out into the pond. Very, very quietly he circled — and the voices called him, so that he moved nearer to them — nearer, afraid but hungry.

He did not see the circle of the net. He knew danger. Then suddenly, from beneath him, came the swift pressure of the moving water and the net all around him, twisting so he could not escape. He twisted and he turned — but he was dragged out of the water. No matter how hard he fought, with his strong tail, his teeth, anything — he was caught and shoved into a dark cage and felt himself being lifted and carried away. He struggled. He heard human voices, not so quiet now but somehow gentle — but no. No. He cringed in the back of the cage.

Terror.

Down in the warm dimness of the lodge, his siblings waited. Close together they waited. Hours passed. They slept. They never awakened.

He did not understand what was happening. Food was thrust into his cage. Because he intended to fight for his life, he ate — aspen, a sweet vegetable almost like a water lily root and something crisp and white he would one day discover as an apple. But he stayed away back in the darkness. During the hours he was being moved he knew only the frightening drone of motors, the chatter of humans and utter, complete loneliness.

At last, after a long time had passed and a long journey had been made, he was alone again. He had a dark kennel filled with clean, warm straw, and a pond, and food. When he was sure no one was near he left the kennel to explore. He could not leave. All around him, too high for climbing, too tough for cutting, was a great wire fence. When a human came near he heard the word Buster often. He ate. He swam. And then he returned to his kennel and stayed in the darkness.

So the days passed. Day after day after day. He only ventured out at night to eat, swim a moment and then back to the safety of that kennel. Day after day after day.

Two hundred miles away, again in a part of our country which had once been a balanced, thriving wilderness but now scratched over with roadways and highways, bridges across which the cars and trucks thundered day and night — even there, in a quiet pond beside a roadway, the beavers had found a place to live. There they had built a lodge, a dam, a pond, and there they lived quietly. Possibly, they had lived there for some years. The mother beaver was large — and since beavers continue to grow in size all their lives, she must have been relatively old. In the darkness of the lodge she was surrounded by her family — possibly kits, and the kits of previous years, and her mate. The family disturbed no one. The humans, who drove the trucks which roared across the bridge did not even know that the beavers were so near.

The younger beavers had been out in the evening, feeding, swimming, playing. Perhaps the parent beavers had been mending the dam; fall was approaching, and the water level had to be set for a comfortable, warm winter in the lodge. However, when the darkness came, the younger beavers dipped beneath the surface of the water and swam down the tunnel to the warmth of their lodge. They had eaten and now they slept — perhaps the parent beavers stayed out a little longer, feeding quietly.

Mother beaver trundled out onto the shore and up the small bank where the grass still grew green and succulent. For awhile she fed and then — she never quite understood. A sound like the crashing of thunder in her ears, and pain — terrible pain — and darkness. Maybe her mate came to her and tried to rouse her — and, when he couldn't, knew that the kits were his responsibility and returned to the lodge. Her darkness lasted a long, long time, and then the pain pulled her back to awareness.

She huddled against the abutment of the bridge. Pain numbed her. She could not move.

One day passed. In the long hours of the night her mate came to her, talking very, very softly — but she could not move and he had to return to the kits. Another day she lay and heard the cars and the trucks and sometimes the trail of harsh radio music as they passed. Until, late in the day, a truck stopped.

She wanted to run. She wanted to fight — to defend herself. But she could not move. A human forced her into a box, and then, darkness, and she knew she was being taken away — far away — away from her family.

This is the story of two beavers, and, as do the stories of most of the animals which come to the Aspen Valley Wildlife Sanctuary, it has begun in tragedy. The telephone call came from the Belleville Humane Society.

"We have an adult beaver just brought in. She has a broken leg.

Do you think you can help her?" Yes. There followed the explanation about finding her by the bridge and the man who brought her in. And her size. "She is huge — but very quiet." We made arrangements to meet the Humane Society van the next day, part way between Rosseau and Belleville — and began to plan about finding her a place at the Sanctuary.

At the junction of highways 169 and 12, I met the van. Here, we have often met and moved animals (raccoons, skunks, foxes, squirrels) from the van to my car — and stayed to chat for a few moments before turning our vehicles around and heading to our homes again. Leslie lifted the cage out of the back of the van and set it on the ground.

My first reaction was simply: "She's huge! — and, she was motionless. I wondered how one would set the back leg of a beaver — perhaps with a pin? Certainly she would chew off any cast.

But that was the veterinarian's worry. I had only to take her home. And face the problem of housing her. A rather difficult problem.

The Sanctuary was almost full. Buster had one of the larger enclosures to himself — a kennel, a pond, plenty of clean straw. And lots of food. Still, most biology books will state that beavers from different colonies, or families, are not compatible. Sometimes they will even kill each other. So, all we could do was to make a high, firm fence down the middle of the enclosure. Buster would have to be content with half. We put the big beaver carefully into the other half, and watched her.

She moved slowly, but she did move. Her leg must have hurt her but she was standing on it and, more importantly the injury seemed to be related to alignment. We did not want to subject her to further stress — we talked to the veterinarian — we left it to heal itself. (Which it has done — perfectly).

We tested the fence carefully, making very certain that it was firm enough and high enough that the beavers would have to stay apart.

Night came — each beaver had lots of food and water and warmth — and we left them to the long hours of darkness.

Early next morning I went out to check that everything was all right with the two beavers. The big beaver was not in her kennel but asleep in the deep straw beside the fence. I presumed that Buster was in his kennel — but — she looked so gigantic! I looked again. Just a little, she stirred. But not all of her.

Cuddled close beside her was Buster. Having not read the biology books he did not know he was supposed to stay away. How he managed to climb the fence I do not know. He did. He had, not his mother — but a mother. And far away, while her mate gave care to her own brood, she gave care to this orphan.

Her care has continued. Months later, they are still together — she teaching him, he learning and not alone. When spring comes, they will go free, together.

Not an altogether happy story, but not altogether sad, either.

COYOTE III

6 – The Remarkably Clever Coyote

"Coyotes: a Death Machine", said the title of the article in the *Toronto Star*, quoting Johnnie the Critter Gitter, who had learned his art from his "granddaddy". He has pictures of what a coyote can do to a sheep — "guts hanging out, necks ripped open, ribs picked clean." And I can't help wondering if he has ever taken the same camera to a slaughter house where humans do the killing.

Whiskey coyote

Never mind. Coyotes do make humans feel defensive — inferior. Because coyotes can outsmart us every time. Not only can they out-manoeuvre, they do it with a gay sense of humour which, when seen, can be only infuriating to us lesser beings.

I love coyotes. I rather think they are more clever — or at least as clever — than the highly respected wolf. At least coyotes have a greater sense of humour.

I have raised both wolves and coyotes and counted it a deep privilege. The wolves grew, learned and were released; we were always aiming in the same direction. With coyotes, the competition is always evident. Yes, one day they would be free — and they would show me how it was done. When the day of release came, the wolves would — after a long siege — be captured and taken to a far away place. Coyotes tend to evade capture for hours and then go over the fence.

Coyotes will be very cooperative as long as they need to be bottle fed. Once weaned, the scheming begins. Yes, they will take the food you give them; yes they will stay in their enclosure — on their terms. A high fence will hold a wolf; a coyote seems to have squirrel genes. They climb. Even a twelve foot fence will not necessarily hold them in. When a coyote knows he can make it on his own — he will go.

Meet Moria. The circumstances were exceptional and we wanted to keep her a second year.

Early in May the weather was warming, birds were starting to come to the trees. At last, at long last, the snow had disappeared — time to release the coyotes who had come the previous year as pups. They were pacing in their enclosure (roofed), sensing all the waking wilderness around them, wanting very much to be free.

Two of the coyotes had come as pups, old enough to be wary of humans but not old enough to go free. Since the coyote enclosure was away across the field, tucked up against the big trees on the hillside, it is quite private, away from human contact. Also, since humans are the greatest enemies of coyotes, we wanted to preserve and encourage this wariness. No one but Tony would approach them and then only to feed them. They had a large, straw-filled shed, underneath which they had burrowed so that they could sleep in the earth. They had branches in which to hide, and a couple of evergreen trees for shade. All they had to do was eat and grow. According to some knowledgeable biologists, more than fifty percent of wild-born coyotes, even with parents, die in their first year. These pups had survived, so we had two coyotes to go in the spring.

Toward the end of the summer, a third coyote came to the Sanctuary from another rehabilitation centre. She had come to them as a small pup, with a violent case of mange. She had been treated thoroughly and was well — and so they sent her here where she would spend the winter with the two other coyotes and become wild again, to be released in the spring. With no problem, the two accepted her. Winter came. Deep cold and deep snow. Three coyotes to go.

However, during that long, long winter another coyote, hit by a speeding car on a busy highway, had a back leg badly broken. So that it would heal straight and strong, a veterinarian inserted a plate. Because this coyote had been so sick, and the bitter winter winds were howling through the valley, we put him in a smaller, warmer enclosure near the barn. He hated it. He hated us. He hated anything to do with captivity. He charged the fence. He hid in his hollow log. We promised him, again and again that when spring came we would set him free. Four coyotes to go.

Spring came — a grey, warm morning in May. We had chosen a far away release place, as far as possible from the coyotes' greatest enemy, the human being. Catching the caged coyote was no great problem, though his fear and resentment were palpable. The three in the enclosure were more of a problem.

Four humans were engaged in the capture — but, though we outnumbered them, they were out-thinking us. When they clawed their way up the high fence, we had to net them, subdue them with heavy blankets and push each into a strong cage. At last — at long last — two of the coyotes were caught. The third stayed, very still and quiet, in the shed.

We should have suspected something.

Capturing her should have been easy with a net across the door. One of the men stepped up to the entrance. Stopped. And listened.

And said: "Pups."

Coyotes are very clever but they do not read. The "Book" had said, with great authority: "Coyotes do not breed until they are two years old."

Ten coyotes to go. But not now. Only three to go now.

To release coyotes anywhere near the habitations of humankind is to expose them (the coyotes) to the worst possible risk. And so our van, with the three caged coyotes, began the long trip from paved road to gravel road, from gravel road to a dirt trail — back into the country where very few humans ever go. Once there, the cages were lifted out, carried even further back, and opened.

The coyotes were away — fast, so fast — no looking back — into the undergrowth and away, back to the wild where they belong. For us, a moment of deep reward. For them? I have often wished that I could understand how that first moment, the moment of realization that they are free, made them feel. The joy and uncertainty of it must surge through their awareness, fiercely, until they are beyond human touch. And then the quiet wonder.

However, back at the sanctuary, still in captivity, was the young mother. Because she was to be with us for one more year, we named her — Moria. Because (we tended, then, to believe biologists as though they were divinely inspired) she had pups, who were not likely to survive their first year — we would feed her, give her privacy and safety.

The enclosure had held coyotes and even timber wolves for several years. The fence had been heightened and kept in good repair.

Tony added heaps of evergreen branches for privacy and allowed no visitors whatsoever. The top of the enclosure was secure. The floor was natural earth but all around the perimeter, for a good six or eight feet, fencing was buried in the ground. Because animals, in trying to dig out, will dig at the edge of the fence, this is a simple way of making escape impossible. It will hold wolves, lions, cougars.

But not Moria. During the hours of the night, she began to dig — not against the fence but out in the middle of the enclosure, beyond the protecting wire spread. She dug in the evergreen, where no-one would notice. She dug a long, long tunnel — out! And, when the eyes of the pups were opening, she carried them, one by one, through her tunnel to freedom.

She lives in one of the caves at the foot of the hillside. Sometimes we hear her at night. We leave food for her. Sometimes she comes and takes it.

No coyotes to go.

Clever, without any doubt. But a sense of humour? Consider this experience with four other coyotes who spent some time at the Sanctuary.

Perhaps their names caused it all: Tequila, Whiskey, Brandy and Rum. My only excuse is that sometimes we get very desperate for names; quite often, unless the personalities of particular creatures are very outstanding, or for a special reason the animal must be differentiated from all the others, we do not attempt to name. Just a justification! This year we had over three hundred raccoons, fifty skunks, over a dozen foxes — good care becomes more important than names.

However, since these coyotes were going to be staying with us for the required year before being given their freedom, we — however unsuitably — named them.

Whiskey, Brandy and Rum were siblings. They had been sent to us from a rehabilitation centre in Toronto; we are very fortunate that we have lots of land and lots of space, so wintering animals is no problem. Tequila had come alone from a Humane Society where she had been loved and somewhat spoiled. A week or so older than the others, she still accepted them quite gleefully, and they all settled into their enclosure, romped and played, ate prodigious amounts and slept.

In the fall, we moved them to the large coyote enclosure up by the hill. The shed, which all the coyotes except Moria had scorned, had been removed except for the wooden box-like platform. It was sturdy enough and we rather thought the coyotes, on fine sunny days, might like to stretch out and enjoy the warmth. Of course, we also hoped that in the relative remoteness of the enclosure, the coyotes would commence going wild.

We fed them, made sure they were clean and healthy, and left them alone. At night they sang their slightly off-key songs. From the distance we could see them romping, digging, tumbling about with each other. Completely happy young animals.

The phone rang.

The caller introduced himself politely. He was a television cameraman from a company in New York State. They had understood that coyotes were very common in Ontario and thus he had been sent north to get some footage of wild coyotes. He had spent several days and had not seen one. No doubt he had been seen by a good many coyotes.

"I understand you have some coyotes in a large pen. Any chance I could get some shots of them?"

The media have always been very supportive of the work we do; cooperation works two ways.

"Yes. We have four young coyotes."

"Is the pen big enough I could get pictures with no fence showing?"

"Depending on the angle, yes".

He came. I took him out to the enclosure. Not a single coyote in sight.

He asked: "Where are they?"

With great confidence I assured him, "Oh — they're there. Hiding in the trees most likely."

He believed me. I believed myself and let him into the pen. Setting his camera at the proper angle, on the wooden platform, took some time. I left him there, closed the gate and locked it (no accidents, please), and went back to do the work that had to be done.

A great silence hung over the valley. One hour. I was busy in the barn. Two hours. Busy out behind the house. Three hours.

I heard him call. I looked up. Camera over his shoulder, he was shaking the gate. Going up, I unlocked it and let him out.

"There isn't a single coyote in that god-damned pen!" he announced, loudly, "not a single god-damned coyote."

"Oh, but," I said, "There is. Tequila and Whiskey, Brandy and Rum."

He snorted. Stomped down to his truck. Got in. Slammed the door. Drove away.

I waited, thought, and closing that gate carefully behind me, went into the enclosure. "Hey guys", I called softly, "Where are you? Tequila? Whiskey? Brandy? Rum?"

I heard the thump of tails. From under the platform.

For three hours the television cameraman had been standing on that platform, standing a few inches above carefully silent coyote pups. I am prepared to swear that they were laughing!

Windwalker

Wolves IV

7 – In the Beginning — Learning About Wolves

I had two errands to do in Huntsville. I had to pick up dog food at the Agro Centre. I had to pick up some medication at the veterinarian's office. I did not know that I was about to begin an adventure which continues to this present day.

While I was waiting for the food at the Agro Centre, a woman came in, also for food. In her arms was a pup, fluffy, bright-eyed, about six weeks old. Of course, I spoke to her.

"She's a wolf-malamute cross", the lady explained, saying nothing more.

I touched the pup, rubbing its ears so that it turned its head and pressed into my hand. Then my dog food had been put into my car. I paid for it and left.

The next stop was at the veterinarian's office. He was busy. I sat in the waiting room — waiting, of course. The woman I had seen with the wolf/dog pup, came into the office with the pup still in her arms.

"Here she is," the woman said, then passed the pup to the receptionist and left — without looking back. The pup was taken away.

When the receptionist returned, I said, "That's a wonderful little pup".

"Yes" she said. "Too bad."

"What do you mean 'too bad'?"

"We're putting her down. She says it has epilepsy".

That was that. The puppy was killed. Mercifully, I am sure, but nevertheless, dead.

I felt sick. As I drove slowly home, I became more and more angry. Epilepsy is treatable. The death was a matter of convenience.

I remembered how the pup had pressed her head into my hand.

A few days later, the phone rang. "There's an advertisement in the Huntsville Forester — two wolf/dog pups for sale at three hundred dollars each."

My anger focused. I made two phone calls. The first resulted in the six hundred dollars needed. The second resulted in a friend volunteering to go to make the actual purchase.

Since then I have come to realize that purchasing the pups merely encourages more irresponsible breeding. The Sanctuary no longer purchases pups. However, when Dave, who had negotiated the purchase and retrieval of the animals, drove up our lane, a pup sitting beside him in the cab, the first wolf/dog arrived at the Sanctuary. Our learning began.

The pup, sitting proudly beside Dave, was beautiful — long-haired and silver grey. There was only one pup. I opened the door. The pup wagged its tail and reached to nuzzle my hand.

"Only one? ", I asked. "What happened to the other?"

Dave motioned under the seat. There she was, squeezed underneath, terrified. I put my hand toward her. She shrank further back into the darkness. She would not come. Both of us had to reach in to drag her out. She trembled in my arms and hid her face against my shoulder.

Dave had just raised a pup he loved dearly. He had had to surrender it to the Seeing Eye Dog School to which it had belonged. In the drive with the pups to the Sanctuary he had fallen in love with the silver pup. I was apprehensive about raising a wolf/cross but I knew what Dave was about to ask.

I anticipated him. "Do you want that pup?" He nodded. I held the frightened one closer to me. She was going to take all the attention I could give her.

Again, these years later, I am not certain that the decision was the correct one. Perhaps, having her friendly brother would have been a help. Hopefully, with experience, one learns. However, Dave took the pup. It has had a good life and is much loved. I was left to do the best I could with the other pup.

Her name is Windwalker.

Windwalker was about ninety percent wolf and certainly the wolf genes in her were dominant.

A wolf is a family-oriented animal who will give lifelong allegiance to the lead wolf, the Alpha wolf. I had little idea of the strength of that bond.

At that time we had an enclosure close to the house. It was about twenty feet long, six feet wide with a shed at one end, a small pond, a large hollow log with lots of evergreen for cover. There Windwalker lived. I visited her regularly, sitting with her, playing with her, feeding her. We became friends — good friends.

Why do adventures come in bunches? Again, the phone rang. Six wild wolf pups had been orphaned. Could we raise them at the Sanctuary? Yes.

The arrival of those pups was, perhaps, the reason I was the one left to become Windwalker's friend. Sanctuary staff learned that six wolf pups require a great deal of care. This was our first experience with raising wolves. The time they absorbed was incredible. The pups became known as the Moonsong Wolves. Windwalker was left in the cage at my door and in my care.

The growth of friendship with her was, at first, slow. She dug herself a den under the hollow log and hid there. If routed out, she bolted into the hay in the shed and cowered down. So, I took to simply sitting in the cage with her, talking to her, waiting. After awhile I could sense that she was listening. Gradually, very gradually, she began to come to me. And then we were friends. Though she was growing steadily, she would come and put her head on my knee and let me rub her ears and under her chin. I never offended her by trying to pick her up.

Knowing wolves are family animals, I kept wondering about her loneliness. Meanwhile, the Moonsong pups had turned on the runt of their litter. In the wild, he would not likely have survived his injuries. He was half their size. They attacked him, ripping at his throat, chewing his legs. He had numerous stitches to close the gaping wounds and so, of course, he had to be separated from his siblings. We wondered if perhaps Windwalker would accept him.

The experiment seemed worth trying. We would watch carefully. We put the little pup in the cage. Windwalker and Moonsong touched noses. Tentatively, tails wagged. Moonsong crouched, inviting play. They chased. They wrestled. They shared food. Tired, they slept together.

However, the friendship lasted only a few days. Whatever flaw had caused the Moonsong siblings to turn on him, began to affect Windwalker. The play turned dangerous for the pup. We had to take him out. Once more, Windwalker was alone.

Because the Moonsong pups were pure wolf, we had decided that, despite no encouragement and a great deal of opposition from authorities, those wolves would be raised to go free. We had to spend time and finances on building them a large, hillside enclosure. Over four hundred feet around, it enclosed pines and aspen, a sandy hillside, a small ravine and a pond. We built warm kennels for them — which they ignored. However, until the wolf pups had gone free, Windwalker had to stay with me.

I enjoyed her. I fed her. I talked to her. I spent time with her. Because she did have dog genes in her, she could not be allowed to go free. I wanted her to like other humans. Still, at the sound of another voice, especially a male voice, she would hide. A few friends tried. Tense, her eyes averted, she began to tolerate them. Merely toleration.

Winter came and passed. Windwalker had grown into a lovely wolf. We accomplished the release of the Moonsong wolves. The large pen was now empty.

Of course, the phone rang. We already had an animal which was ninety percent wolf: a wonderful creature with a name which was quite romantic — Windwalker, echoing the mystery of Ghost Riders in the Sky. And now — the phone rang. We were about to acquire another wolf/dog — named Bill. Bill!

The phone call was from the Bracebridge Humane Society. "We have just had a dog brought in — except he is not a dog. He is part wolf". Then, without a pause, a little defensively, "He's quite beautiful. And very friendly. But we can't adopt him out, of course." So?

"He's been neutered — and he's easy to handle, and —" I heard myself saying that, of course, we would take him. A friend for Windwalker? I hoped so. As it happened, the possibility that he was either a sibling or a close relative of Windwalker's was very strong. We learned more about him when the family who had tried to have him for a pet visited him at the Sanctuary.

Bill had come from Aspdin — a very, very small community less than ten miles from the Sanctuary. Windwalker had come from the same place. And, when Bill was put into the enclosure with Windwalker, recognition seemed to occur. Sniffing, stiff legs, still tails — and then, quite suddenly, play.

Windwalker was playing.

By the time Bill arrived at the Sanctuary, the Moonsong wolves had been released and the big enclosure was empty. We had cleaned it, repaired it, strengthened it. We now gave it to Windwalker and Bill.

When his first family visited Bill, he greeted them much as a dog would — leaping against the fence, wiggling, barking. He then left them to go to romp with Windwalker.

"We couldn't keep him," the people explained, quite honestly. "He wasn't like a dog. They told us he would be like a dog but he wasn't. We couldn't leave him alone. He would destroy the house. He chewed furniture and doors. He walked through screened windows and screened doors. We've obedience trained dogs before but he wouldn't be trained. We couldn't confine him. We just couldn't keep him."

Bill looked like a beautiful malamute. But he was also wolf. He was much more friendly than Windwalker. If a visitor came to the fence, he would bounce over, put his nose through, wag his tail — dog. Except, sometimes, for reasons only he knew, he would take a great dislike to a specific person. Windwalker stayed back in the trees, watching from a distance, never trusting, not wanting to be seen — wolf.

Only once have I ever had such opposition to a "dangerous" animal arriving at the Sanctuary. That was when Teka, a declawed cougar came and word spread (as it always does!) through the neighbourhood. I began to get threatening calls on the telephone.

"Have you got a cougar there?"

"Yes". (I actually thought someone else might be as pleased as I was.)

"Well get rid of that damned cat or I'll shoot it." A pause. "And you, too"

Windwalker and Bill caused a similar reaction.

Back then our road was simply a gravel country road — dead-end. It did not even have a name. Traffic was minimal. The new wolf enclosure was thirty feet back from the road, screened by high bushes and the row of evergreens which bordered the road. Unless one was aware of its existence, and even then having to look closely, one would never see it. However, a couple of local "heroes" learned about the wolves. These men I had encountered on a number of occasions during the hunting season on nearby Bear Cave Road. I did not admire them. They certainly did not admire me! The first time I heard the gunshots right out in front of the house, I was puzzled — after all, it was not hunting season. An old truck was driving by — and again the next day, and the next. The "hero" began to make a habit of driving up and down the hill by the wolf enclosure, firing his gun out of the window of his truck. I knew who he was, of course — everyone knows everyone in a community like ours.

As long as he did not harm the wolves, I rather shrugged my shoulders at his antics. He took to driving by at night, and firing his gun. I phoned the police. "Get the license number", they said. Sure! Go out long after dark to confront an angry man who was, most likely, at that time of the night — drunk! I was not so sure I wanted such an encounter.

After awhile the continual harassment stopped, flaring up only occasionally — I suspect usually when the "hero" was drunk enough. Still, a drunk man with a gun is no joke.

Windwalker and Bill did well in their pen. Windwalker had lots of privacy. Bill enjoyed whatever life presented him. Since I was busy with tasks related to the growing Sanctuary, they were in the care of two people who loved wolves living across the road from the Sanctuary, but, at the time, working with the Sanctuary.

Over the next few years the Sanctuary grew — beavers and bears, deer and moose, dozens of raccoons and squirrels, foxes and coyotes, birds too. We were busy caring for them, building enclosures, carrying out the education programme, recording all the facts which would help us to learn about raising and releasing wolves — much too busy. Our staff was enlarged. And all the time I did not have time to think much about Windwalker. I knew she was being cared for. I forgot she was a wolf.

Over those years I saw Windwalker from time to time. I heard about her. She was shy, she would come to some women, cautiously. When I did have free time, much of it was taken with walking my dogs, one of which was a Pit Bull. Always on a leash, we walked up the road every afternoon. I knew where to look under the evergreens and through the bushes to see Windwalker, and call her. She always came to the fence, watching, tail moving back and forth.

I wished I could take her out to walk her — but she was a wolf. Huge and beautiful but a wolf.

Then came the horrendous winter of the wolves — a story to be told later. The couple who had been looking after Bill and Windwalker told me that they would love to continue caring for them on their property, since we would be so crowded with new wolves, at the Sanctuary. Under no consideration! Bill and Windwalker would move to a new enclosure further back on Sanctuary property.

This enclosure would be built amongst the thick, young pines on our hillside. Another seven thousand dollars! A volunteer organized all that. Workmen were difficult to find, somewhat unwilling and always behind schedule. However, the day eventually came and the enclosure was finished. An influx of bear cubs caused further delay and then the difficulty of getting the right tranquilizer. After that we had to coordinate the veterinarian's schedule with ours. Finally, finally everything came together. Ian, the veterinarian, prepared the tranquilizer. A volunteer loaded and used the dart gun.

Bill's move was as easy as it could possibly be; he was darted, went to sleep, was moved quickly and woke up in his new pen. Windwalker's move was altogether different. She hid in the trees.
She would not allow the volunteer to approach her close enough to dart her. When he did succeed, her adrenaline was flowing so fast that it countered the effect of the drug. Moving her took almost four hours. Finally, in the new pen with Bill hovering near and anxious, she took a long time to awaken.

The new enclosure was around a big stand of young, thick pines, their branches low to the ground. The morning after the move, on early inspection, Windwalker was nowhere to be found. Not in the enclosure. Bill was there, frisky, happy — but no Windwalker.

Concerned, I crossed the valley to the enclosure. I walked around the far side toward the back, away from other humans. I knelt down to try to see under the pines. I called.

Windwalker charged out from under the pines. She danced against the fence. She licked my face and my hands. She whirled round and round. She was home. After seven years she was home. I felt ashamed. I had not known.

I promised her that never again would I leave her for so long. My knees were getting older and the snow came — deep snow. I couldn't get across every day to see her. But I could call and she would answer — and she knew, and now I knew that I was her Alpha/human and she was my wolf. An infinite privilege.

Tony Grant came to be the manager of the Sanctuary. Bill was almost instantly his friend. I think the dog genes were strong enough in Bill that he could relate to humans far more easily than Windwalker ever did. She was not afraid of Tony. She would wait nearby when he took food into her — but she was aloof. I saw her as often as I could. She would dance to the fence, and we had good conversations.

Windwalker had a couple of good years and then she began to lose weight. She moved as little as possible. She stopped eating. We decided that, one way or another, she had to be seen by a veterinarian. As in the years before, catching her was a problem. This time, feeling rather like a traitor, I called her to me. Tony was able to tranquilize her. We put her into the van and she was taken to the veterinarian's office.

Mouth cancer. No cure. When they put her to sleep I could think only that, at last she was free, a whole wolf, running and running . . . .

Only a year later, Bill, too, contracted cancer. Sometimes I wonder whether this disease is another gift we humans are giving to the animals we hold captive.

8 – The Winter of the Wolves

The problems with humans and the wolves was, all of a sudden, so severe that I actually began to speculate on some sort of spiritual conflict — wolves can have that effect on people! Humans react strongly to them: They either hate wolves intensely or love them intensely. Neither state is necessarily rooted in either knowledge or wisdom. Wolves, highly intelligent, react. At any rate, tension surrounding the wolves that winter (1994) was palpable.

One of the manifestations of that love-hate relationship between humans and wolves is the obsession many humans have to "own" a wolf. Even the concept that such ownership is possible shows an abysmal ignorance of the nature of the proud and independent wolf. Once, in an effort to raise an awareness of the extent of this ownership problem, I made a poster for a local event where we had been invited to set up a booth. To catch the attention of passers-by, the heading on the poster was, "Wolf Pups for Sale". The poster went on to explain the foolishness of such a situation. However, few people read beyond the headline. Even in a small place like our village, I could have sold a dozen pups. "Owning" a wolf sounds glamorous.

Until wolf pups are no longer available, somewhere, somehow, a Sanctuary for captive-bred wolves and wolf/dogs ought to be established. Accustomed to humans, and having no family of their own and therefore outcasts in the wild, they cannot be released. The present alternative?

They can live out their lives in tiny enclosures or at the end of a chain. They can be killed. Or (and such a place does not exist.) they can do well in a Sanctuary built for them — with huge, wooded enclosures, complete privacy and as natural a life as possible.

Best of all, we should have enforceable legislation prohibiting the breeding of captive wolves.

Still, at the beginning of the 21st century, we are being continually asked to give a home to captive wolves. The Sanctuary is full. We cannot respond to the need.

Brandy and Kuhuli, Heidi and Brandy II

In the spring of 1994 I was dealing with two very persistent couples who had wolves in desperate need of sanctuary. The first indication of what was to come was a call from the Midland Wildlife Centre of the Ontario Humane Society. They had been approached several times concerning a pair of Arctic Wolves. The young couple who were caring for them had acquired them in an act of compassion. They had been born to a pair of wolves who belonged to a wayside zoo near Barrie. The adult wolves were being given temporary shelter by the young people because the zoo was overcrowded. The wolves gave birth to two pups. When the adult wolves were returned to the zoo, the pups were not wanted there so the couple kept them, trying to save them from entering the buy-and-sell life of zoo livestock. Their own domestic dog acted as their mother. With that dog, the wolf pups formed a strong bond; perhaps they regarded her as the Alpha wolf. For several years things went well but then, the dog died.

The wolf hierarchy was broken. The personalities of the wolves seemed to change. Certainly, no matter what care they received from humans, they were not happy.

Some time later, a neighbour's child put her hand into the enclosure. One of the wolves grabbed it and held on. The adults had to beat the wolf's nose until the nose bled and the wolf was forced to release the hand of the child. The child had to be treated at the Sick Children's Hospital in Toronto.

That wolf was neither fierce nor wicked; it was responding to an intrusion in a perfectly normal way. The child was not part of its pack; she was invading their territory. She was small and therefore prey. The child most likely thought she saw a big white dog; she approached a large white wolf. However, finding a home for the wolves became imperative. We had no room. I had to refuse.

Meanwhile, the young couple commandeered some of their friends, and arrived here in assorted trucks, with assorted tools, wearing clothes that indicated their intention for hard work. And the work was, indeed, hard. It took all day. But at the end, they had reconstructed the wolf enclosure. Out by the side of the hill, this enclosure was surrounded by a grove of young evergreens, some meadow and a muddy spot — wolves do like some mud! And then, the next morning, they brought the two beautiful Arctic Wolves — Brandy and Kuhula.

Almost immediately, another series of persistent phone calls began. Two five year old wolves, having been purchased from a zoo as pups, were now living in downtown Toronto. Their only place for exercise was on a tar roof-top. Though the wolves were taken into the house at night, they broke through closed doors, showed aggressive behaviour toward some people and toward three hybrid wolves who shared the house with them. Because they were so strong and aggressive, they could no longer be walked on a leash. Again, I told the woman who called that we could not take them. We did not have the money to build another large enclosure. She continued to phone. She asked how much it would cost to build the enclosure. Underestimating grossly, I said it would be about two thousand dollars. She promised the money. The removal of the wolves from Toronto was urgent. Finally, I said I would get a temporary enclosure ready.

The young people with the Arctic wolves had not given up either. They suggested that they take down the enclosure in which the wolves were presently living and re-erect it in another location here. They would do all the work themselves. They would move the wolves themselves. They would continue to bring food to them. And over the years they have been true to their word.

The couple in Toronto were also ready to make promises. Though the wolves accepted both of them, they were becoming more and more aggressive. The phone calls became daily. They promised the two thousand dollars toward the construction of the pen. Their insistence became so extreme, I finally said that we would convert one of the smaller enclosures into a temporary place for their wolves. They were afraid even that would take too long.

Barry Young (on the Sanctuary part-time staff) secured the help of a few volunteers to raise the height of the fences. However, the pressure to bring these two wolves to the Sanctuary was so intense that I finally agreed to confine two of our resident wolves (Treena and Sundance) to a small part of their cement enclosure. We were still trying to build a good place for them. Four wolves would be in a small enclosure not big enough for a fox. The couple assured Mike, one of our board members, that the money would be coming quickly.

So the Toronto couple arrived with the two wolves in the back seat of their car. The trailer, which they pulled behind the car had been intended as travel accommodation for the wolves. Before they were out of Toronto, the wolves had torn a hole in it. To keep the wolves quiet on the trip, they had fed them cheeseburgers. Now leashed, the two wolves were taken to the cement enclosure.

One wolf, Heidi, had what is considered typical wolf markings. Brandy, who had been referred to as a black wolf was actually a beautiful silver grey. When I asked why they had referred to him as a black wolf, the explanation given was that, as a pup, he had been black. Over the years the colour had changed. This happens.

The woman explained that one reason she wanted the wolves out of the house was because of her husband's condition. He had had a nervous breakdown. She intimated all sorts of family problems. They had spent time in British Columbia; had moved to Toronto to run a coffee shop.

She talked a good deal and seemed to be very nervous. The man was quite tall, a grey, stony looking man with a single pig tail down to his waist. He also talked continually.

We had put evergreen branches to form a screen between the two sets of wolves, though the branches were soon pulled down. Fighting was never a problem. The Arctic wolves were very quiet, relaxed. Those newly acquired from Toronto were extremely tense.

Barry and his volunteers worked quickly raising the height of the fence on the hybrid enclosure. When it was finally ready, because the two wolves were still so aggressive, we asked the couple from Toronto to come to help us move them. They came. When they tried to walk the wolves on leashes to the new enclosure, it was very obvious why they could not be walked in Toronto. They jumped. They fought. I was relieved when they were safely inside the enclosure.

When they were released inside, the wolves, noses down to the ground, began to explore the long grass, the real earth, the pine trees — a tiny touch of wilderness which should have been theirs.
The husband expressed concern about the strength of the wire on the gate. though it had held a pair of hybrids for a good number of years. So, Barry doubled the wire. We gave the wolves an extra large meal. The Toronto couple finally left. The wolves howled.

Barry enjoyed the wolves and spent a lot of time with them. The Arctic wolves exhibited acceptance, then playfulness, even a sense of humour. The timber wolves were tense, suspicious.

One of them showed a limited desire for a restricted friendship. The Toronto couple continued to visit. They always brought, not good food, but "treats" for the wolves. They always entered the enclosure. The wolves would, naturally, welcome them. At first I did not monitor the visits closely. However, after each visit, the wolves would be very upset. Their distress spread to the other wolves and the whole community would be in some distress for several days. Though the timber wolves seemed to enjoy the actual visits, the parting was causing deep distress — and not allowing us to make much progress in getting the wolves to settle happily. Barry was working to make the wolves trust him. Their lives could be content only if they could trust us. Trust was coming, but very slowly. Another visit from the Toronto couple would destroy all the progress we had made.

The couple had made great promises about helping us to build a big enclosure. The two thousand dollars never materialized. First they had to put air-conditioning in their restaurant.

Then the family needed some sports equipment. The man said he was trying to sell his piano, but couldn't because there was a key which would not work. Would we take the trailer — with the hole in it which the wolves had made — and try to sell it? In total they gave us two hundred dollars toward the enclosure. The visits, the ensuing disturbances continued.

After one such visit which left the timber wolves especially stressed, we felt it would be better for the wolves, though upsetting to the couple, if the visits were restricted. We put chains and locks on all the wolves' enclosures.

I had begun to dread the regular visits — first of all for the sake of the wolves and then for my own sake. The woman always seemed to be nervous, watching my reaction to her husband's peculiarities, defensive but pathetically loyal to him. Sometimes I felt that she was afraid of him. Once he said they were looking for property up this way. When I told him of a couple of available spaces, he reacted sharply; he certainly did not want to deal with any real estate agent.

Another time he told me how lucky I was to live in the country — city vibrations were so upsetting, making him feel turbulent. Once a loon, calling, flew overhead. The man told me that he, himself, was a loon. He clammered back at the overhead loon, so noisily that none of us could enjoy the bird. Then he said he understood wolves because he was part wolf himself. A very disturbed man — still, my concern was for the wolves. If they were to have quiet lives, the visits had to be restricted.

At the end of a long, hot day, as I stepped out of the side cages on the barn after the last check of the raccoons and foxes, I saw the couple, once again coming up my driveway. My heart sank. I knew I had to talk to them about the tumult their frequent visits caused. And, the pens were locked.

Carefully, reasonably, politely, I tried to explain. The wolves bonded to them, were being forced to go through the separation and grieving process again and again, destroying all the progress we were making with them. As we talked, Heidi and Brandy yelped. The man kept howling back at them, increasing the tension. He was furious. His wife backed him up. I still have the impression that he would be very abusive if she didn't.

First — if he could not go into the pen, neither could Barry. The wolves were his. They must not bond with anyone else. They belonged to him. He owned them. He had not given them to us. Then the attack turned personal. I was an old woman who thought she knew everything but really knew nothing at all. I lived away back in the country and was ignorant about life in general.

Because I was extremely tired, I sat down on the grass to listen to him rage on. Sitting down was a mistake. He towered over me and went on and on and on. Finally, seeing no end to it, I suggested that he talk to Barry — Barry can be calm and reasonable. He didn't want to talk to Barry. I got up and went to the telephone anyway. When I returned, they had gone.

The man had been extremely abusive and I was frightened. Both Barry and Mike felt that I should call the police. I did!

The night seemed quiet. My dogs did not stir. I heard wolves howling but that was quite usual.

Next morning, when I looked out toward Heidi and Brandy's enclosure, I saw that the gate was open. The couple had come in the night, cut the chain and stolen the two wolves away.

Treena and Sundance

The laws respecting wildlife are rather indefinite — and certainly enforced only when the Ministry is directly challenged or happens to have a spare officer with some time on his hands. Technically, owning a wild animal is illegal; technically, keeping a native wild animal in captivity is illegal. Therefore, an animal, purchased from a zoo is not considered to be a wild animal. Thus, some zoos are willing to sell wolf pups to humans, who are dazzled with the perceived glamour of owning a real wolf; unfortunately, the market is not small and the money is good.

 The source of wolf pups is the breeder who dances through a hole in the law by obtaining his breeding stock from outside the province. His wolves are not native, not indigenous. He may sell them freely. The humans who want to "own" a wolf have little idea of the responsibilities involved. Wolves are not dogs!

 Only the wolves themselves know that, whatever their origins, be it privately bred or zoo bred, they are born to be free. Every instinct is honed to wander silently, to hunt, to belong (with a lifetime commitment) to a pack — to live its own life in the wilderness.

 A Wild Animal Kingdom, in eastern Ontario, had wolves for sale. Treena was a white Arctic wolf whose ancestors came from the North West Territories and so she was unprotected by Ontario law. For three years, alone, she paced around and around a circular cement-floored cage; she did not have even a small kennel in which to hide. Sundance, another white wolf whose ancestors were all zoo born, was at the same zoo, enclosed in a four foot by six foot pen, no den, no shelter and — according to a sign on the fence, for sale.

 Our friend, Shirley, who saw and objected, approached the owner: "Why do you sell wolf pups — who buys them?"

 He laughed. "I sell 'em to 'wops' in Toronto. They think they make good guard dogs." Legal.

Shirley phoned us. She told us about the two wolves. When Shirley is rescuing an animal, no discussion is necessary. "If I get them, will you take them?"

"Yes, Shirley." We would have taken them whoever asked but I am certain I sounded just a bit meeker when the question was asked by Shirley, whom I like, and admire.

We built a large enclosure deep in the woods, quiet and private. Because they had been stared at by humans quite enough we wanted them where they would not have to encounter any humans except their keeper — but we made one mistake. The mistake was not that the wolves should have quiet lives but our choice of location. Perhaps we had much to learn from the wolves themselves — wolves do not generally trust human beings.

The Sanctuary property was, in one place, divided by fifteen acres of private land. At the time the owners were enthusiastic about the wolf enclosure location — not only would they work with us in allowing us to cross their property, but they would feed the wolves and care for them. All seemed well.

When everything was prepared for the coming of the wolves, a few of our volunteer young people drove down to the zoo to collect the wolves and bring them home to the Sanctuary. They took a camera with them. We have pictures of the cage where Treena lived. I was told the story of how she had the spunk to fight the men who were her keepers as they tried to force her into a cage. How could she understand what new, brutal treatment might be in store for her? Her keeper took the back of a shovel to hit her and drive her into the cage. She cringed, snarling in the back corner.

However, through a winter snowstorm, she was driven to her new home in the great stillness of the wilderness.

So, for three years, their new keeper brought them meat in a bucket and left it for them. They were not necessarily friendly nor were they aggressive. When she scattered the meat on the ground, they stood back and waited until she had left and the quiet once more closed in. They ate and grew fat. Though they could not be free because of their previous life with humans, they seemed to be content.

And then, to use a highly unoriginal phrase: "all hell broke loose".

I was in my house working. All was quiet. Then — a pounding at the door and hysterical screaming. The keeper of the wolves.

"They have my dog! They've killed my dog!"

Outside, I urged her into the car. We drove up the hill, out behind the house and down the laneway as far as we could and ran across to the pen. Sure enough, out under the low pines, a small black bundle of fur. A dead dog. And pacing wolves.

Wolves are wolves and humans must never for a moment forget it. The little dog had trailed the keeper everywhere. She even allowed it to go back with her when she fed the wolves. Of course, she did not deliberately take it into the enclosure — but neither did she remember to shut the gate properly. The dog, quite instinctively, followed her in — and the wolves, with equal instinct, saw live prey. Neither was the dog nor the wolves to blame.

The keeper was still hysterical. We must retrieve the body of her dog! We must! Because her agitation was so loud and uncontrolled, I could see that those pacing wolves were watching us. I hesitated to go in. She threatened to climb the fence. She threatened to go in herself and I continued to wonder how the wolves would react to her extreme emotion. I was alone with her. She ran back to her house and came out with a gun. I took it. She continued sobbing. I suggested she go to the place where she worked to see if she could get someone to help us.

Barry, our wolf specialist, was at work. I did not know anyone else in the area who understood wolves. But asking for help turned out to be a second mistake.

While I waited, the wolves were intensely aware of my presence. I did not want the body of the dog mutilated so when they approached it, I fired the gun into the air. By the time the keeper had arrived with her help, the wolves had retreated almost out of sight, away back in the evergreens at the far side of the enclosure. When, shouting continually and carrying the loaded gun, the helper went in to get the dead dog, the wolves stayed as far away from him as it was possible for them to be — watching and silent. He carried the dog out. I shut and locked the gate.

The keeper wailed a little less loudly. Her friend informed me (as though ordered to by the Almighty), "I have to tell you that human life is more valuable than these animals. You have no right to keep such dangerous animals." I was polite. I made no answer. After all, he had just retrieved the dead dog from the wolf pen.

I thought that, although the event had been tragic enough, it was over. Barry took over the job of feeding the wolves. Over? I had not taken the keeper's two sons into account. Both sons have quite lengthy police records, and, with a certain regularity, end up in court and in jail. Strutting, macho and, always in their contacts with me, they were slightly arrogant and condescending. I realize that most of their attitude problems came from a family background of confusion, rejection and indulgence. Now they had a chance to be "heroes". They could rise to the defense of their mother!

They decided their mother had been attacked by the wolves. They took her to a doctor. He found a faint scratch on one leg, which may or may not have been caused by an animal. There was absolutely no evidence of a wolf attack. The sons, nevertheless, were determined to be "heroes".

They did not have the courage to approach me. They phoned a member of the Sanctuary board, described the vicious attack and demanded that the wolves be killed. Demanded! The wolves were vicious! Vicious! Vicious! They had to die!

When the board member phoned me and I heard that the wolves were being accused of attacking, I was angry. I denied it. Quite rightly, he was worried about law suits, insurance and the complete disintegration of the Sanctuary. I was afraid and quite alone. The sons contacted the friend from the mother's work-place who was very willing to come and shoot the wolves. I did not know how to fight the death sentence.

The sons were boasting around the neighbourhood, posturing, the grand protectors of their little mother.

A friend had come to work with me. She stayed at the Sanctuary while I, though aware of the looming sentence of death, had to make the trip into Parry Sound to pick up the food for all the other animals. I know I drove too fast, but even so, as I turned onto our gravel road, I passed an ATV (All Terrain Vehicle), trailer on behind, our workplace friend, gun in hand, driving. To kill Treena and Sundance. As I wheeled my car into our lane, Chris, my friend was there. We watched the ATV go up the hill and turn into the lane.

"They are going to kill the wolves", I said.

No!

No. Chris leapt into the car, we sped up the hill and around the circular drive so that we ended up between the ATV and the lane back to the enclosure. The ATV had to stop.

He faced us, gun in hand.

"No", we said, "you are not going to kill them"

"I have permission", he said.

We did not care — nor answer. How things would work out we did not know. Only that Treena and Sundance would not die. We did not move.

He waited. A heavy, belligerent silence. Finally he backed out. "I'll have you know I'm worried about rabies", he said with his best Almighty God look. "That dead dog was covered with saliva. I have my family to think about". He turned the ATV and left.

The second son announced that we had two days to get the wolves moved out or they would kill them themselves.

Moving semi-wild wolves is more difficult than moving dogs — one cannot whistle, say "come!" and snap on a leash. Tranquilizer for dogs does not necessarily work on wolves. A special tranquilizer, available only through a veterinarian, had to be sent up from the United States. Because Ian, our veterinarian, cared about wolves, he ordered it immediately. Still, there was no way the drug would arrive in time. Another board member talked to the sons. They finally set a new dead line. Two days. And the drug did not arrive.

Sunday. Son number two charged up the driveway in his car, braked hard and strutted to my door. Confident, finally, he would face me. We had only until that afternoon to remove the wolves. I told him we were doing our best — the drug had not arrived. He didn't care. He would shoot the wolves himself. Then he flung himself back into the car and drove away.

The wolves belonged to the Sanctuary. Their enclosure was on our property. Unfortunately, the only reasonable access to it, was not. If the sons killed the wolves, they would be committing a crime — but they were "heroic", they would go to jail to save their mother.

I told him I was phoning the police.

He said I would be sorry. He would phone his friends in a motorcycle gang.

Barry and Chris went up the hill to talk to the mother. She had such a fit of hysterics I could hear her wailing all the way down to my house. She was, she insisted, unable to control her sons. Barry and Chris returned. In a little while the silence was broken by the drone of motorcycle motors. Down the hill, past our lane and up the hill — bike after bike. Twenty-six motorcycles.

I phoned the police.

They came. They were here for several hours. They wore bullet proof vests. And they warned me to stay in the house. We waited. The police went up the hill. We waited. A long, long time passed. All they had gained was time: until Wednesday. The gallant sons were ready to go to jail.

I could imagine them posturing bravely in front of their tough friends.

But the drug was still delayed at the border.

To express the wonder, the dignity, the mystery of a wolf is impossible. They are aloof, aware of a meaning, a worthiness to life which is outside our understanding. I was furious that the lives of two of these magnificent creatures should be at the mercy of two difficult humans.

Wolves do not kill to pander to inflated egos, or to prove their worth.

One son was due to arrive on Wednesday, gun in hand, to kill. The only pen we had for the wolves was a 10 by 20 cement enclosure — too much like the ones from which we had originally rescued them.

The drugs crossed the border. They would arrive on Thursday.

"Wednesday", said the son. "I will shoot them Wednesday at noon."

Once again we called the police. The police visited the mother at work. She insisted she could not prevent the murder. They told her that her son would go to jail. Once more; hysterics. Finally she said that, if we could guarantee to move the wolves the moment the drugs arrived, she would lie to her sons. When they phoned on Wednesday morning she would tell them the wolves were gone.

Gallant fellows to reduce their mother to that; sad comment that she would lie! However, the dysfunctional family was not my problem. The wolves were.

Finally, on Thursday, the drugs did arrive. The veterinarian, Barry and Chris spent hours in the enclosure. The wolves, quite aware of the extreme tension, were difficult to approach near enough to dart.

But, at last, the wolves were moved into the little cement enclosure. The only improvement over the zoo was that they did have a dark, private, kennel.

We built a new enclosure, away back on our own side of the road. Trees, meadow, a pond — where Treena and Sundance lived quietly and well. Some years later, Treena, grown quite old, died. Sundance, now in the company of a lively little coyote named, Silver, lives there still.

9 – Amarook, the Wolf from Way Up North

Often, as evening darkens across the valley, the wolves begin to sing — at first one, and then another — until the valley is brimming with song. Wolf song. Infinitely beautiful.

More than thirty years ago, when first I moved to this valley, I wrote about the wild wolves who moved through our woodland. Raccoons and beavers, deer and foxes had already found sanctuary here. But not wolves. Searching for something else, through a pile of old papers, I found something that I wrote way back then:

> I know I should like wolves. I have read *Never Cry Wolf* and, putting the book down, have been full of admiration for the wolf, anger at a stupid government and a desire to write as well as Farley Mowatt, the author. One evening after I had read the book, I was sitting on an old wooden bench behind the original log house. I was listening to the singing of a pack of wolves in the swamp at the bottom of the pasture. The late evening sun slanted across the meadows, the logs at my back were warm from the sun of the summer day. My dog lay close to my feet. The wolf song was haunting and romantic.

In November when darkness closes in early, the fields and woods were dank with long rain and cold. Huddled in one corner of the barn was Bandit, an apartment raised raccoon with one leg so injured I did not know if he could go free. He was desperately afraid of the new life into which he was being introduced. Especially of dogs. I had shut them in the house so he would not be disturbed as I sat with him in the darkness and tried to win his confidence.

And then the wolves began to howl.

The wolves were at the pasture's edge, just beyond the pond. The pack was in full cry, shrill and wild, shredding the quiet of the evening.

Bandit cowered in his corner. The deer in their pen lay unmoving, tense. I went to the barn door and listened as the sound filled the valley, echoing from the rocks, engulfing us all in the frantic, lusting hunt. I wondered about the doe and her two fawns who often grazed at the edge of the meadow. I wondered about Willow, our deer who was soon to be released. I knew about the stern code of survival of the fittest. About live and let live. About the right of all things — even predators — to live. Still, I felt the terrible stillness of the raccoon and of the deer.

From the ridge, the torrent of violent song went on and on.

"Go away," I willed them, "Go far, far away."

From far, far away one howl answered them. I felt fear as one does when thunder rolls across the hills and lightning jags into the valley. One can do nothing except wait it out.

Finally the tumult subsided. Silence flowed back across the valley. The deer remained tense. Bandit would not allow me to touch him. But, for that night, the wolves were gone.

I thought, "if someone gave me a wolf pup and said, 'look after this', I would. I would give it all the love and care I give to this raccoon, to this deer." But the fear did not go away.

That was written, as I said, thirty years ago. Since then, many wolves have come to the Sanctuary. Nightly, I hear their song.

And one song, high and clear, is the voice of Amarook. She should be singing, not across a quiet valley in Muskoka, but over the cold, barren lands of the north. She should be — perhaps, in her song, she is calling to those places — I don't know.

About a hundred years ago (an approximation — but sometimes it seems that long!), when I was a grade seven student at Alexandra School in St.Catharines, Ontario, we had a very excellent teacher — Miss Hannahson — who enjoyed history. We enjoyed it, too, even though we had to memorize all the routes of the early explorers and be able to trace them out on maps of Canada as well as spell their names correctly. We were taught the journey of Henry Hudson, up the eastern coast of Canada, through the Hudson Straight, past Ungava Bay, into Hudson Bay, where his crew mutinied and where he met a rather ghastly end. Thus, when I received a phone call from "up near Ungava Bay", I did know where the call came from, more or less — though I had never heard of Kuujjuaq.

I wonder if Henry Hudson had heard of Kuujjuaq.

In Muskoka the September sun was warm and gold. In Kuujjuaq, a blizzard was raging.

The voice, over the static of his cell phone, said: "We need a home for a young wolf. Four months old. Could you take her?"

Yes. The abundance of static and blizzard prevented much conversation. Later we learned the whole story of the pup whose name, Amarook, was the Inuit word for wolf..

I did not know if Kuujjuaq existed when Henry Hudson sailed by, but I am quite sure that if it did, things there have changed considerably. First of all, the call came from a cell phone — a fairly new invention. The call came from a hunt camp — another fairly recent innovation. Lastly, the gentleman making the call offered to fly the wolf pup to us in his own private plane — definitely a mode of transport unavailable to Henry. All that aside, I agreed to give the wolf a home.

Her story is rather sad. Kuujjuaq seems to be situated in tundra country, not so Arctic that grey wolves cannot live there, but certainly sufficiently Arctic that a blizzard would be raging in early September. Sometime in the spring, Inuit children had found a wolf den; why it was vulnerable to their predacious activities we do not know. They found three tiny wolf cubs, eyes not yet open, and they took them home. After a while, the novelty wore off or, at least, the children lost interest — the pups were passed to another family. Two of them died. One pup was not much responsibility but even so, she too was given away. So her life began.

Her eyes had opened. She was unpredictable — not exactly dog, not fully understood as a wolf. She did not really belong to anyone (wolves never do). Sometimes she was allowed to run free. Sometimes she was chained. She was fed with the sled dogs. Then, again for a reason we do not know, she was given to the men at the hunt camp. Here, at least, she enjoyed the company of the hunting dogs. She ran with them. She was fed with them. She slept with them.

She became very, very tame.

The camp men were good to her. Her attitude was (and still is) that life is a joyous affair, people are good and all is well with the world. If she had been correct, all of us would have a wonderful time. However, the men knew that when the goose hunt was over and they had to return to their homes in the United States, they could not leave her alone. She had grown to a good-sized animal, obviously a wolf. If she sought out other humans (as she undoubtedly would), she would be shot.

And so, the phone call to the Sanctuary.

"We will be flying down to Montreal tomorrow. Could you meet us at the airport?"

No. However, Harriette, from the Urban Wildlife Centre which does excellent work against all odds in downtown Montreal, was prepared for almost anything. I phoned her. She always replies to phone calls in French — unfortunately, I have to reply in English. Without missing a beat, she switches languages. "A wolf? From Kuujjuaq? At the airport tomorrow? Certainly."

Nothing fazes Harriette. "We'll keep it until you get here."

Next morning, here in Muskoka where the sun was still warm and bright and even the trees had barely begun to turn colours, we had another call from Kuujjuaq — they were still in the midst of an intense blizzard and could not fly. Three days were to go by before we had to go to Montreal to retrieve the wolf.

So Amarook came to us.

At four months old, she was as large as a German Shepherd. Her coat was silver — maybe because she had been well cared for, almost shining. Her yellow eyes danced with friendship. She seemed to love everyone. All wonderful — but, had she been set free into the wild, it would be almost surely fatal.

At first, as we were preparing a large enclosure for her and were looking for a companion, (wolves do not like to be alone) she had to remain in a small enclosure near the lane. We knew she would be spending her life with us because of her previous association with people; therefore we made great effort to be friends with her — not difficult! Like a growing puppy, she loved to have her ears or her belly rubbed, to play tug of war with anything handy. Like a growing wolf, she eyed the passing cats and the squirrels outside her pen, and the raccoons in a nearby pen, with nothing playful in her intentions.

Janet, one of our staff members, undertook her care. Quickly, Amarook realized that Janet was her very special friend.

Amarook grew. Today she is a huge silver wolf, living in a five acre enclosure with a wolf/dog whom she allows to dominate her. Depending on her mood, she hides back in the trees or romps to the fence for attention. These years later, while all humans are interesting and therefore tolerated, only Janet knows her well enough, and has gained her trust enough, that we can be certain that she and only she can enter the enclosure in complete safety, feed, talk and be with Amarook for awhile.

Amarook is still a wolf. At night, when I hear her sing, her voice high and clear above the other wolves, I remember that she should be singing across the tundra in the far north. Though she is having a good life, she is loved and she loves, she is not at home.

Maybe her family, all those years ago, sang across the snows and Henry Hudson stopped to listen.

10 – The Christmas Wolves

One year, on the day before Christmas, two young wolves were brought to us. They had been confiscated from a wolf dealer who had been intending to breed them, then raise and sell their pups. Unfortunately for him, but fortunately for the wolves, he did not know about the out-of-province restriction, or else he did not expect it to be enforced. A Conservation Officer had delivered the wolves to a Humane Society. Tony collected them. With the van doors opened after their arrival at the Sanctuary, I saw the wolves crouching in their cages.

The male was tall and black, his eyes yellow, wary, unmoving. In the cage beside him, the female crouched. Slightly smaller, white but with those same yellow, wary eyes, she watched every move we made — unmoving. Janet brought the ATV, with a trailer on behind.

Carefully, she and Tony lifted the cages over and then, driving slowly, they took the long trail back to a seven acre enclosure, deep in the woods, where a stream tumbled through in a series of small waterfalls, into a pool. Quite different from a backyard in a large city.

As quietly as though they had spent all their lives in the wild, the wolves vanished into the trees.

All we saw was the trail in the snow where they had gone; all we heard was deep silence. What were they thinking? No motor roar. No whistles. No horns. No sirens. A deep, deep quiet. I wondered if that quietness flowed over them, and if they understood that this was what wild wolves knew that this was what life should have had for them. Would, if we had any choice, would have for them still.

The officials had promised us that we could take the wolves north, and set them free. And then a court case began. The breeder was suing the official the case was going to court. We could not free the wolves — not yet. A year later, while humans argue and delay, the wolves are waiting. We could not free the wolves — not yet. A year later, while humans argue and delay, the wolves are still waiting.

To share our homes with dogs and cats is a privilege which we have and enjoy — the creatures as comfortable and happy as the humans. Wild animals cannot ever be owned without a spirit somehow broken, a potential never realized. And yet humans persist. Perhaps the feeling of power — of subduing the wild — is still strong. And it is still wrong.

More than any other large animal, in this country, humans seem to want to own wolves.

Maybe it is a reaction to the long held belief that wolves were ravaging beasts, dangerous, powerful how great the human who can bend it to his will. A good market for wolf pups does exist and is legal, as long as the original wolf stock comes from outside our province. The advertisement: "Wolf pups for sale" is a real money maker.

These Christmas wolves do not like people. If they see a human approaching, they hide. From somewhere amongst the trees they wait, very, very still, until they are alone again. Even when Janet takes them food, she does not linger. These wolves have chosen privacy. We hope to give them freedom.

Tonight, as every night, the valley quiet will be broken when the wolves begin to sing. First, the Arctic wolves and then the high voice of Amarook. Sundance, who began her life in a cement enclosure in a zoo, will lift her nose and sing. One by one, others will begin. And presently, from their enclosure far back in the woods, the two new wolves will join — high, clear, beautiful.

Perhaps next winter they will sing from a frozen lake — free — many, many miles away from any human.

11 – Once Upon a Time There Was a Wolf
LOKI

The man's voice on the telephone was jocular, macho — I didn't like it. But I cared about the wolf he was offering — an eight year old Minnesota Red Wolf that needed a home.

"I offered him to the Ministry of Natural Resources, but they didn't want him," he announced. That, in itself, sounded strange to me but still I listened. "Got to find a home for him soon." When I asked why the urgency for a new home was so sudden, he didn't really explain.

Is it pure wolf?" I asked.

"Of course," he answered. "Couldn't be purer".

A few days later, Janet drove over to pick up the wolf. And so Loki came to the Sanctuary.

Wolf? Mostly, according to the papers which came with him: bred by an irresponsible man in British Columbia who claimed he was saving the Minnesota Red Wolf from extinction. Certainly some dog blood — perhaps collie — ran in his veins. The papers proudly claimed that his grandfather was the wolf used in the film *The Bear*. But the animal who lay on the floor of the van was huge, red, and very, very sick.

"If I take him for walks, he runs away" — the voice on the telephone had explained. This wolf could barely stand. His shoulders hunched. His hips dragged low. His head hung. His eyes were as lifeless as those of a zombie. For eight years he had lived on the end of a twelve foot chain. A wolf, who will roam for hundreds of miles, held captive on the end of a twelve foot chain for eight years!
Not illegal. The wolf was not from Ontario.

We gave him a warm, straw-filled stall in the barn.

He began his life at Aspen Valley. First of all, good food, as much exercise as he could handle — and a warm, warm place to sleep. At first his walks were merely a little bit outside of the barn and then he would collapse. He ate little. His dark eyes seemed to be fixed someplace far away — on the hills and valleys he had never known?

Gradually, gradually, he began to be able to walk further, even to show some interest in life. He ate more. When someone came near his enclosure, he would lift his head. The day when he greeted Janet with a wag of his tail was a day of celebration

Loki was given medication. We thought his arthritis was easing. Occasionally he could walk a good way along the road — even around the beaver pond and back. When the warm weather finally came, he could spend his days outside, sleeping in the sunshine, even watching the activity around him — cats, the occasional skunk, and ducks and humans. Sometimes he stood and wandered a bit. Another red-letter day came when he made it as far as the enclosure where my house dogs were, and sniffed at them tentatively. They seemed somewhat puzzled by him because he did not respond to their enthusiastic barking.

Still, he did not put on weight and he ate little.

What sort of person would allow an animal to reach this condition? Eight years on the end of a chain was bad enough — the untruths he had told us so that we would take the wolf were indefensible. That he should be unaware of the condition to which the wolf had been reduced was impossible. That he should have boasted about owning a wolf — well, humanity is not always the most compassionate of creatures.

We finally took Loki to the veterinarian for the gift of a merciful death. The last few days he could not eat. He could not walk. He had to be carried to the van on a stretcher. I believe that Loki is now free to run the wild hills of a new and wonderful life.

And I wonder, perhaps somewhat bitterly, when we as humans will do the basic minimum for such creatures. At the least, it should be illegal to breed, sell and keep wild creatures for our own pleasure.

12 – A Cello In the Woods

THE MISSOURI WOLVES

Dark hills, dark trees, stars and a campfire; cello music, soft and sweet, tangling with the murmur of the wind in the stars; a tent, a young man playing his cello — and two wolf pups, listening. Alex and the pups are alone in the woods.

Alex was very different. The world of young people swirled in throbbing noise around the cities. Young bikers, back packs, boots, canoes, followed the trails, trodden and untrodden, deep into the wilderness. Alex followed those trails, and, finding a remote valley, pitched his tent and stayed. With two wolf pups and his cello. Different.

An advertisement had appeared in a newspaper in Missouri. "Pure bred wolf pups for sale." Alex was angry that anyone would presume to own a wolf — to buy it, to keep it in captivity — wolf, meant to roam the deep wilderness that Alex loved; and then — an idea! The pups were expensive. Alex had little money. He scraped together every cent he had. When that wasn't enough, he badgered his friends. Then he bought the two wolf pups — black pup and a grey pup. When he went back to his tent in the woods, cello on his back, the little pups followed him.

These were very lucky pups. Though they had been sold for money and thus parted from their mother (she would later die a captive), Alex vowed to teach them to live in the wild. The day would come — he vowed it would come — when they would be independent, free wolves.

He would be the "parent wolf". Though the process would go very much against his own instincts against killing, he had to trap and kill small rodents and birds and bring them back to the camp site to cut them up for the pups to eat. A parent wolf would, of course, eat the prey and then regurgitate it for the pups. Alex was rather glad he was incapable of that! As the pups grew, Alex had to catch rabbits and grouse, woodchucks, raccoons — any small game available. The pups, following, had to learn.

Alex learned from the wolves. He talked to them. He knew what they wanted and they, very surely, knew what he wanted. The communication was not necessarily in words: thoughts, feelings, strong urges — no one word describes the process. Mental telepathy sounds, somehow, artificial. Only the human who has experienced that deep communication with an animal can really understand it. But it happens and to know it is a great privilege.

The same mystery surrounds the response to music. I have found that some animals respond to some music. Deep, unrelenting, throbbing sounds can be very disturbing. Gentle, quiet music can be soothing, even inviting.

The cubs grew to know the sound of the music, the sound of Alex's voice; they became able to understand the unspoken thoughts of his mind. They grew; they followed him when he hunted for their food and they hunted on their own. Watching them, Alex knew that the day would come when they would be able to be self-sufficient wolves, living freely in the wilderness.

But not in Tennessee. Wolves were hunted ruthlessly, ignorantly, there. So, where? He spread out the maps. He looked for a wide stretch of unpopulated country, as free as possible from human danger, with a good supply of food and few other wolves. The Rockies, of course.

And so, again, he gathered what little money he could, an old truck and headed west. He knew he would be able to do what biologists are now spending thousands of dollars attempting to do — reintroduce wolves where they have been obliterated. However, Alex did not have a university degree in biology. He did not know that he was pioneering a new adventure. He did know whether the wolves would survive.

So, the wolves following, he hiked up a mountain trail, heading as far away as he could, treading the trail he had plotted to the remote vastness of a distant valley. The wolves were eager. And then, one night, the wolves did not come to the campfire. He could see them, out in the shadows — wary. A Conservation Officer appeared. "What are you trying to do?"

Alex was unwise enough to attempt to explain.

"By whose authority are you releasing wolves?"

His own.

"What university is behind you?"

None.

"Are the wolves healthy?"

"Yes. Very. And able to fend for themselves." The wolves, trusting no human except Alex, faded into the night time.

"You may not release them. You must return with them to Tennessee and keep them captive. For always. The fine for releasing them will be five thousand dollars, jail for you and death for them." Alex took the wolves back to the campsite in Tennessee.

If I were a cartoonist I would make a drawing of the border between Canada and the United States, a straight line cut through miles of standing trees, over hills, through valleys, across lakes. In the trees, on each side of the line, would be a wolf, a male wolf and a female wolf, looking at each other with love-filled eyes, tears streaming down their noses. "We can't," they would sob, "we mustn't mix American and Canadian DNA." The forestry departments of both nations would applaud.

In a secret sequel, those wolves would cross the border, mate, then produce a family which would be fine with them but might mix-up the careful records of government ministry.

Alex was determined these wolves would be free. His every inquiry met with total rejection. If, at this point, the music from his cello became very melancholy, that melancholy was completely understandable. However, a network of concerned humans does most certainly exist.

One evening the phone at the Sanctuary rang. I answered. Introducing himself, Alex explained his situation. I listened.

He did not explain about the cello.

However, over the long distance telephone we began to make plans for the wolves. Down in Tennessee, he spread out the maps of Canada. Here, outside the little Ontario village, we did the same. We searched again — for a very remote location, lots of food, few other wolves and no conservation officers.

We found a place.

And, we planned carefully. If we "sinned" against the biologists, we did not sin against the Creator or the wolves. Actually, they crossed the United States/Canadian border as Huskies.

The north shore of Lake Superior can be very stormy. The day that Alex and the wolves, with a friend of the Sanctuary, in a covered boat, made their way along that shore it was heavy with fog — and rough. The great grey waves heaved against the tall grey rocks of the shoreline. The map which was in use now was topographical; they knew exactly where they were heading. And they found the spot — a cove, sheltered by two arms of huge rocks and hidden by evergreens. The water within it was still. On a small beach of sand, they landed the boat.

Alex opened the cage door. The wolves were out. Free.

Alex stayed with them only a little while, talking to them, telling them that behind them stretched a vast country, where they could live as wolves are meant to live. He sat on a rock, with his eyes closed and his head bent on his knee, and, soundlessly, he told them.

As the boat drew away from the shore, across the quiet water of the cove, the wolves stood on the grey rocks and lifted their muzzles — and they sang . . . .

Before this story — this true story — can be fully understood, two important facts about wolves must be known.

First, wolves are intensely family oriented. Separating the pups from their mother and then from each other, is only the beginning of the travesty inherent in the breeding of captive wolves. To wolves, their offspring, their siblings, their parents — even aunts and uncles — are of prime importance. These relationships are not only intense, they are lifelong. Broken from his family at an early age, the wolf pup has lost his frame of reference for life. To the pup, the emptiness may be undefinable, but it is very real.

Secondly, animals communicate in many ways. We humans see

them barking, snarling, whining, growling, howling — sounds that some creatures have in wide variation.. These we hear and interpret in our own context. However, another level of communication does exist. Some humans may call it either extra-sensory perception or telepathy. It may be those or it may be much more. It can be an inner speaking, as defined as actual words would be between humans but soundless. It can cover great distances. This communication is not a "talk to the animals" in a Disney fantasy nor is it an anthropomorphic projection about what I might think the dog I adore says to me. This communication is deep, meaningful — and sometimes almost frightening.

So, back to the story of the four wolves.

Timber and Midnight

For the human concerned with the next two wolves, I will use the innocuous name of Smith. Because this is a true story, and because the law did become involved, anonymity seems wise. Since he deals in the trade of exotic animals, Smith saw an advertisement in a newspaper in Tennessee: Pure Bred Wolf Pups for sale. Great! Smith could add two more novelty creatures to his collection of captives. He already had a cougar and a monkey. Easy money could be made in buying and selling exotic animals. And wolves were in demand. Lots of humans want to own wolves.

Smith made the trip to Tennessee. He had money in his pocket, quite a bit of it. He bought two small pups (a grey pup and a black pup, one male, one female), loaded them into a kennel in the back of his car and drove the long, hot highways back to Ontario.

Smith lived in a large city, in a tall house with a small, fenced backyard — a part of which he was willing to share with the wolves. He called in a construction crew. They put down a concrete pad, twenty feet by twenty feet. Concrete is easy to keep clean. Walls of concrete blocks (neighbours would not be able to see through those!), rose twelve feet on all sides. He filled a kennel with hay, gave them a big dish for water — what more could a wolf want?

He fed them canned dog food. They ate and grew fat.

Smith took great delight in his wolves. He invited friends to visit them. He boasted that the wolves loved him. They came when he called — crawling on their bellies, submissive.

Some day, he let it be known, they would be old enough to breed and his friends (or whoever) could buy and own wolves, too.

The black pup and the grey pup grew. On the hard concrete the pads of their feet grew calloused. They paced, twenty feet this way, twenty feet back. In the hot sun, they lay close to whatever shade the cement block walls could give them. In rain and snow they stayed out in the weather (maybe enjoying it? — maybe knowing this was part of the life they should have had?). They heard the blare of the constantly playing music, the whine of passing traffic, humans talking, laughing, dogs barking — the unceasing day and night-throb of a busy city.

And, perhaps, as a sign of promise, in that small square of sky above them, in the long, long nights, they saw the far and silent stars.

Then Smith's world began to crumble — just a little. The city where Smith lived and kept his wolves passed a by-law, forbidding the buying, keeping or selling of exotic pets. The fact he had a cougar, a monkey and two wolves was very well known — he much enjoyed the publicity. He decided to ignore the law. He moved the cougar and the monkey away. He had no place for the wolves to go.

He tried to keep the wolves quiet. He tried to pretend he did not understand the law. But the official notice was handed to him: he had ten days to get rid of the wolves or the grey wolf and the black wolf would be confiscated and destroyed.

Smith heard of Aspen Valley Wildlife Sanctuary. Now, I rather suspect, he thought we would be easily manipulated. He phoned.

"If they come here", I said, "they become our wolves and we will do all we can to return them to the wild."

Glibly (why didn't I suspect?), he agreed. Yes, yes, yes — but would we please take his wolves before the city shot them?

At the time, the fact that we did not have an empty facility for the wolves seemed merely inconvenient. We did not know it would be the one fact which ultimately saved their lives. While we immediately began the construction of a large wolf enclosure, the wolves had to be taken to the barns of the Ontario Humane Society in Newmarket. For that privilege, Smith had to sign papers that the wolves were the legal property of the Humane Society.

We knew when the wolves were to be given to the Humane Society. Because Smith had wailed to the local newspapers, the event was given television coverage. We watched. If the picture was designed to arouse pity, and possibly support for the owner losing his wonderful pets, the coverage was a failure. When Smith approached the wolves, they cowered away from him. He had to put tight nooses around their necks and drag them to the Society van. Whatever else we might have to do to prepare them for the wild, we would not have to break any bonding to the human race.

Our first responsibility would be to give the wolves some idea of what life in the wild would be. For their enclosure we chose a location well back in the woods, far away from roads or any other human-caused noise. The enclosure would be as large as we could afford (we pestered many friends for money!) and, as natural,. In the end, however, the enclosure was only four hundred feet around. (Today our wolf enclosures are each several acres). It enclosed a dozen good sized pines, a few aspen, alder and a small pond. The ground was warm and soft with pine needles. All around the fence, the trees of the forest pressed close. Aware that the wolves were living in an indoor pen, and were waiting, we built the enclosure as fast as we could.

Once, I went down to the Humane Society barn to visit them. Clean, well fed, and warm, they were very wary of humans. They stayed as far back in the cage as they could, silent and watching. They showed not one sign of trust in or desire for human companionship. I wanted to tell them that, after a time, they would be free. At present between us was a barrier of fear.

Meanwhile, at the Sanctuary, destroying as little of the natural woods as was possible, we were working quickly. The fence line was cleared. The post holes were dug, and then the posts were planted and made firm. Twelve foot fencing was raised all around. Because wolves dig, the fencing had to go underground. To ensure that the wolves would not escape into the woods here, we also laid fencing on the ground, around the perimeter of the fencing and fastened it secure. No amount of digging would give them freedom. For that they would have to wait for a much safer place than Muskoka. Though to us, and perhaps the wolves, the construction seemed to take forever, we actually managed to complete it in only two weeks. We were ready. We sent word down to the Humane Society.

Once more the wolves were caught, caged and loaded into the van. Once more they made the two hour drive over noisy, busy highways. But this time, when the van inched its way through the trees to the gate of the enclosure, the wolves would begin a life they had never known. That night, as they experienced the real woods for the first time, what did they think?

- soft pine needles beneath their feet
- pure water from a spring in the ground
- night birds calling — an owl
- in the distance, wolves

Their food changed, too. Now, real meat — even, when road kill was available, deer, grouse, rabbit. They could move freely and a little further. Soon, their fat began to give over to muscle.

At last they had solitude. Their only human contact was when one person, silently, brought them food. No attempt was made to coax them to be friendly. They were left to the stillness of the woods.

We named the grey wolf Timber and the black wolf Midnight.

The wolves had been living in the enclosure only a week or so before the calls from Smith began. At first I gave him credit for being really concerned. His voice had just the correct amount of pathos.

"How are my wolves? I miss them!"

"They're doing very well".

"Do my wolves miss me?"

An unfortunate tendency to complete honesty made my answer somewhat brutal.

Perhaps the possessive "my" should have warned me. His calls continued. His voice gained in pathos. Then, one day, he announced, a note of triumph faint beneath the pathos, "I miss them so much. I'm building a pen outside the city limits. I want them back."

No. I had promised the wolves that they would live free. I would keep that promise.

The phone calls stopped. A notice to appear in court arrived.

The Humane Society inspector was not alarmed. He smiled a little as he showed me the papers. "Smith has given us legal custody of the wolves," he said. "The wolves are under our jurisdiction. Don't worry". Nevertheless, I did.

The date was set. Along with the representatives from the Ontario Humane Society, we were to appear before an Animal Care Review Board. It would decide whether Smith could reclaim his wolves. Even as I drove out the winding, hilly gravel road in Muskoka, where the Sanctuary is located, I was determined that these wolves would never be returned to their captive lives. Regardless of any ruling, those wolves would be free — if I had to return and open the gates and simply let them go (even their chances here in Muskoka would be better than lifelong imprisonment) — I would do that. And I would pay —

I would pay, not the Sanctuary — any fines involved. Even a term in jail, I reasoned rather bravely, would be shorter than the times the wolves might have to spend in captivity. At any rate, my thoughts were muddled perhaps, but I could not forget the black wolf and the grey wolf, back in the shadows of the pine trees, capable of freedom.

And so the trip to the city. I had grown up in a city. I had lived for several years in the chaos of New York city. Now, however, as I drove with the Humane Society people, down the crowded highways, the speeding cars and the noise, I think I understood a little of what Midnight and Timber must feel. The invasion of noise, movement — all unnaturalness. Standing, waiting for a few moments, outside the towering building where the panel awaited us, I was bitterly homesick for the quiet, gravel road in Muskoka.

We were seated in a small room, facing three men who sat behind a long desk on a raised dais. Through the wide windows behind them, spread a great expanse of roof-tops, chimneys, smoke.

The room still seemed very dim.

For the first time I saw Smith — tall, blonde, slim, elegant. He sat between his lawyer and an attractive young woman, who (I learned later) was a stripper. We were in the row of seats just behind them. I could not help noticing that the woman, who sat with her ankles crossed and her toes on the floor, thus exhibiting the soles of her shoes, was wearing new ones from which she had not removed the price tag. I expect the fact that I could notice and be amused by such triviality was some indication of the fact that I would find the entire situation intimidating.

Smith was called upon first. He was questioned about the wolves; where they had come from; what care he had given them; what he intended to do with them. His performance was quite good. He told how much he loved them, how they loved him. He cared for them. He understood them. He was building them a grand, new enclosure outside the city.

The officials from the Humane Society were questioned. The inspector presented the papers which Smith had signed, legally giving the wolves to the Humane Society.

"Why are they, then, at the Aspen Valley Wildlife Sanctuary"?

"Because the Sanctuary has the facilities to care for them, the knowledge and experience to do what is best for them. They are still under our jurisdiction."

That statement could have been the end of the case; the legality of it could not be disputed. Still, I was called to answer questions.

"What does the Sanctuary intend to do with the wolves?"

"Give them their freedom."

"Would they be able to survive?"

"Their chances would be good."

"How do you know that?"

"We have prepared wolves for release several times. And released them."

"Successfully?"

"Yes."

"Where?"

I was purposefully vague. I would not betray wolves, already free, to any threat from humans. "In carefully chosen places, with good food supplies and few other wolves. Most important, in places utterly remote from human habitation."

The long silences between the questions seemed more ominous than the questions themselves.

After a time, when the three men were consulting amongst themselves and we waited, one turned to me suddenly and demanded brusquely, "If you had to build a cage how big would it be?"

I don't know why I was suddenly angry. I said, "No cage is big enough for wolves!"

The man looked at me for a moment and slowly nodded.

He nodded. The wolves would go free.

That evening, as I was leaving the big city behind and driving home to the relative quiet of Muskoka, to the tall pines, the rocks and the lakes — I wanted to sing. However, since singing is for me almost impossible — I stopped and wrote. The poetry is not grand, but . . .

> No cage is big enough for wolves
> Theirs to run the wide wilderness
> the high hills and deep valleys
> in wind wild snows
> and pine-warmed summer breeze
> under star strewn skies
> under storm torn clouds
>
> Theirs to sleep by the stillness of predawn lakes
> or deep under drifting snow
> in forest dens
>
> Theirs to hunt on shadowy pathways through the trees
> Theirs to play with pups on sunny summer slopes
>
> Theirs to sing from high rock ridges
> to hear answer from a far valley
> moonlight over all
>
> For wolves
> No cage is big enough.

The weeks went by. We fed the wolves. We gave them their privacy. We began the search for the best release site. Then a letter came from Alex.

In the letter was included papers which showed the pedigree of his wolves — the wolves which were now living free north of Lake

Superior. The name of the breeder, the location, the name of the female wolf and the male who had been the parents. The number of pups in the litter and the date of their birth. All very interesting.

A few days later another letter came; this, the papers the Humane Society had obtained from Smith. Again — the name of the breeder, the location in Tennessee, the name of the female wolf and the male who had been the parents, the number of the pups in the litter and the date of their birth.

Identical. Siblings.

Reunion!

I phoned Alex. Family is very important to wolves.

We did not need to search any further for a release area; the cove on the northern shore of the Great Lake, where the first two wolves had gone just weeks earlier, was the obvious place. Arrangements went ahead quickly. Alex met two of our staff at an airport in the north. They would wait there, meet the wolves who would be sent up by plane and then, taking the boat once more along the rocky shoreline, find the place where the first two wolves had gone.

The plan was good and eventually worked. At the appointed time, they were at the northern airport. On schedule, we here managed to lure Timber and Midnight into cages sufficiently secure, and small, to be taken on the plane. The plane, a small bush plane, was hired and waiting. Waiting for a series of strong summer thunder storms to end. The storms were not precisely here nor were they at the northern airport — just every place in between. We had no way of contacting Alex to explain the delay. After two days, I had a phone call saying they had given up and thinking something drastic must have happened were heading here. I told them to stay where they were. The weather would clear and finally it did.

Timber and Midnight waited by the lane, crouched in their crates, apprehensive. The truck which was to take them to the dock and the airplane was late. I was anxious. I walked by Timber's cage. And stopped, as though I had been called — though I hadn't. No sound. I looked at Timber. He was standing up, looking at me. In his eyes was an awareness and intense intelligence, a silent voice — so strong it had stopped me. He was not saying "thank you" but he was silently speaking. He understood how we had fought for him. He understood. That was all. I think that, if some day we meet again, we will be friends.

The truck came; the airplane flew them north. There they were met and taken on the little boat along the rough and rocky coast to the cove. And there they were released. Alex stayed awhile, on the rocks, listening — waiting.

When I talked to him again, he said, "They have met".

"Four voices," he said.

He stayed awhile, locked in by the storms, knowing that four wolves were close but wild now.

One day the storms seemed to quieten. He thought he should leave the wolves to their own lives. He started out along the shore toward the distant docks. The storms were not over. He'd lost the boat, the motor, all his belongings, all his supplies. Except the cello. He saved that.

"Oops" at rest.

Bears

V

13 – The Story of the Three Bears

The phone call was from Detroit, Michigan. The caller identified himself as a policeman. Instantly, I felt guilty and defensive; how could I possibly be in trouble in Detroit? I had never once in my life been in Detroit! When he explained, I breathed deeply and relaxed. Two bears. To be here permanently.

Outside, the valley was green and quiet in the early spring sunshine. Most of the wild creatures from the previous year had been released — the first little ones — tiny, hairless squirrels, unweaned raccoon kits — were just beginning to arrive. Another few weeks and we would once again be unbelievably busy. And now — not cubs — but two adult bears — who could not ever be released.

Since the first days of the Sanctuary, we had raised many, many bear cubs. For the first sixteen to eighteen months of their lives they had lived here, waiting until they were old enough to be taken far into the wilderness and released to live free lives. Years and years ago, when this was an abandoned farm on a dirt road, wild bears had come by. I could remember walking that rarely traveled, winding road, enjoying the wilderness, watching my two dogs ranging along the edges, knowing that we were far away from the busyness of cars and other people. We walked down the hill, around the curve by the tall old spruce trees which are still standing here, and looked out across the overgrown pasture. Away at the back, standing on his hind legs, watching us, was a black bear — just for a moment Then he lowered himself so we could not see him for the tall weeds and grasses. He had been free. The bears who were now to come to make the valley their home, would never know that freedom.

Now, the telephone conversation over, I sat down on a lawn chair and looked over the valley and wondered what I had actually agreed to, and if we could actually manage it — or afford it. I knew I had, without consulting a single soul, committed the Sanctuary to the expenditure of thousands of dollars we didn't have.

And I thought about Detroit — that city to which I have never been — and I visualized skyscrapers, tall shadowy and cliff-like, minuscule cars crawling along the streets which twisted at their feet, blaring horns, sirens, crowds and crowds of people, subways shaking the earth. A city. The heart of the city surrounded by grey slums and sprawling suburbs. Always movement, noisy, throbbing. Housing developments stretching crooked fingers out into the countryside; one arm hooking around an old farmyard; a dilapidated barn, a dark, dark stable: where two bears had been confined for seven years.

Perhaps Detroit is different from the other cities I have known. At any rate, in one corner of an old building in those concrete "cliffs" is a small office of the Anti-Cruelty Society which is aware of some of the horror perpetrated not only on children but on dogs and cats and birds in the big city — and bears. Walking those busy streets is one policeman, who, while he must deal daily with all the contingencies of those streets, still has time to be concerned about two bears imprisoned in a dark barn.

The bears belonged to a group of men who wanted every advantage when they went hunting; important as a good gun were a few well trained dogs. And they could train not only their own dogs, but the dogs of other hunters as well setting up a sort of business. Good idea! All they needed was a couple of bear cubs — not hard to come by.

From a zoo, where the cubs could be purchased, or from a mother bear who could be easily shot, they obtained two bear cubs. They strengthened the walls of the stall. They got lots of hay and straw and left-over doughnuts and sweets from the bakeries. Not much of a lay-out. In the barnyard they built an enclosure with high, unclimbable walls — and they were in business. They could put the cubs in the enclosure, then turn the dogs on them — and the dogs would learn what it was like to hunt bears. Almost in business — the men did not want their dogs to be hurt so the bear's claws were pulled out of their paws and their teeth were cut out.

Both cubs were Black Bears: the male black in colour and the female a light brown. The colour variation is relatively common in Black Bears. With some degree of originality, the men named their cubs, Black and Red. For seven long years, Black and Red were taken from their stall only when there were dogs to be trained. For seven years, the handicapped bears defended themselves the best they could. The dogs learned to hunt. The men made money.

How the policeman and the man from the Anti-Cruelty Society found the bears I do not know. Nor am I exactly sure how they knew about the Aspen Valley Wildlife Sanctuary. But here, in our green, green valley, the telephone rang.

Over the days which followed, with telephone conversations back and forth, I learned more about Black and Red. The bears, mutilated as they were, could not live free in the woods. They could not climb. Berry-gathering without claws or teeth would be very difficult. They had never been properly fed; would that lead to problems in bone development as they grew older? Had life in a dark stall impeded muscle growth? Black and Red were seven years old — bears can live to be thirty or forty years of age. They would be well beyond my life expectancy! I would have to make a commitment for the younger people who were taking over the care of the Sanctuary. Still, the answer remained: "Yes" For better or worse, Black and Red would become part of the Aspen Valley Wildlife Sanctuary.

The only place we had for them, immediately ready, was the enclosure on the west side of the barn, perhaps thirty feet long, fifteen feet wide, cement floored, firmly roofed.. Though better than a dark stable, it was not ideal for bears. We knew we would have to find the money to build an enclosure of some acres — but for now, because the need was immediate, Black and Red would have to live beside the barn. We made the enclosure as pleasant as we could: deep in pine needles, evergreen branches to hide in, a wooden den. And a bathtub, which happened to be bright yellow (donations come as they come!), to do duty as a pond. It was fed from above, through the fenced roof, by a bright green hose. When the next phone call came — the bears were on their way! — we filled great bowls of dog kibble, apples, grapes — all the fruit we could find — and the bathtub with cool clean water. And settled down to wait.

June 5 was a warm summer day. About eight o'clock in the morning the final phone call came.

The bears were indeed on their way. They had just cleared Customs at Sault Ste. Marie. While we worked, cleaning, feeding all the other creatures, we thought about those bears. They were coming along highway 17, one of the most beautiful in all of Canada — across the north of Georgian Bay, through the great rocks and forests. The next call was from Sudbury; they had turned south onto highway 69 — still through huge, smooth rocks, trees windblown by the storms of Georgian Bay. We notified the people who were to come and help us move the bears. Around four o'clock in the afternoon, the call was from Parry Sound half an hour away. Four of us stood on that rise by the barn doors and waited and watched the clock and the road. Then we heard the motor and saw the van crest the hill, come carefully down the long curve into the valley, vanish for a moment behind the evergreens, and then — the U-haul crawled up the lane and stopped at the foot of the rise.

The U-haul had a huge lobster pictured on the side. We didn't have time then, but, since, rather wondered if we could somehow discover a connection; so far we haven't.

The back doors of the van were opened but all we could see in the darkness was one cage and another behind it. The bears were huddled black shadows — unmoving. We would have to carry the cages one by one.

They backed the van as close as they could to the door of the enclosure, which meant that the van was at a most fearful angle. I felt somewhat panicky — bears in cages tumbling down a hillside just in advance of a rolling van — but the men did not seem unduly concerned. They lifted Black's cage out, opened the door slowly, out of the van and into the enclosure. Then Red was moved, again slowly, out and in. It was as though the bears didn't really care what was happening now. Life held no interest for them. They stood, heads lowered, waiting for whatever was to happen next.

And, for the first time, we could clearly see the bears. Both were small for their age and very, very skinny. Black had very little hair — we thought of him as a leather bear. What hair he did have was in small clumps and dry and rusty. Red was even smaller. She had no overcoat — her undercoat was dull and sparse. In great bare patches on her neck we could see long scars, both were dull eyed, lethargic, uninterested.

Until they saw the food. We had gathered everything we could find: kibble, eggs, apples, lettuce, nectarines, chicken. They sat down and ate. (I can't help making a very politically incorrect observation. Black, the male, sat on the food. Red had to get what she could from the edges).

And afterward, when at long last evening began to spread across the valley, they pressed their noses against the chain link fence, snuffing in the green, leafy fragrance of the springtime, hearing the birds, the running water, the wind.

"Soon", we promised, "you will have a huge enclosure with trees and rocks and a meadow and water. You will live there and never be harmed again."

Humans had been cruel to them. We were certain that, once the plight of the bears was known, hundreds of people would be willing to help give them a good life.

The naming of an animal, especially an animal which is going to be part of the family for the rest of its — and your — life, is a major responsibility. We felt that "Black" and "Red" didn't somehow (to use a slang phrase) "cut it". Meanwhile, we were meditating on the challenge, and giving the bears all the help we could to turn them into healthy looking bears.

We were walking the valley, trying to decide the best location for a large permanent enclosure. We wanted a bit of meadow, a stream where we could dam a pond, lots and lots of trees — though none, at least near where the fence would be, that they might possibly, in a surge of strength and energy, climb.

While we were working on the future, the bears were working on the present: eating. They rolled the hollow log around as though it were a toy. They trashed the lovely evergreen branches. Their favourite employment was the yellow bathtub, full of water. Black would sit in it, on his back, his growing big belly above the water line. Until he finished, Red would pace, waiting her turn. One day, we knew they were well on their way to becoming real bears. Ron, one of our volunteers, had turned on the hose and filled the tub with clean water. Quietly, Red was watching . . . and, while the tub filled, Ron was not paying close attention. Quietly, quietly, Red reached , took the hose in her front paws and turned it directly on Ron. He nearly drowned! After a moment, she dropped the hose and walked away. And came back to wallow in the tub of water.

But the names. Black and Red. Traditionally, to some Ojibway, the bear is a sacred animal — not, unfortunately, to all. However, we did have an Ojibway friend who heard we had the bears, and came to visit.

"Their names?", he asked.

"Red and Black".

For one long moment he looked at me and then turned away. "No", he said. "I will tell you their real names." He walked off down the slope to the bear enclosure. Well over an hour later he returned. "The black one is named, Nungoon,", he said. "It means, 'All the Stars'. And the brown one is Giisus. That means, 'The Sun."

So, Nungoon and Giisus they became. While Nungoon and Giisus (our friend told us to pronounce the latter name carefully and clearly, so that people did not think we had named him, "Jesus") ate their way to health, basked in the sun, and scented the wonderful wild, we set about raising enough money to build their enclosure.

Our world is a very needy place: starvation, homelessness, natural disasters and wars — there is no end to the demands on human generosity and goodness. Could we expect that enough people would care about two mutilated bears? Yes — we could. And did.

At first we simply told our friends. The first gift came from those who know a great deal about animal abuse — The Parry Sound Humane Society. The word spread.

The manager of Johnson's Wax Summer Resort for employees, near Rosseau, heard about the bears and convinced his company to make a substantial donation. That donation was duly reported in the local newspaper, the Parry Sound North Star, and went on the news wires. The papers from Huntsville and Bracebridge picked it up and then, the television stations from North Bay, through Barrie to Toronto. The *Toronto Sun* and the *Globe and Mail* covered it. Since the story was on the press wire, it turned up in newspapers from Nova Scotia to British Columbia. Hundreds of people cared and sent donations. We were amazed and overwhelmed, and convinced that, though there were humans like those men down in the old barn in Detroit, who had hurt the bears so badly, there were many, many more who knew the value of the lives of two handicapped bears, and did something about it.

Winter was coming and we knew we did not have time to build

the large enclosure. Instead, under Tony's capable management, we built a smaller place, deep in the pine trees, where the bears could hibernate. Though it was only twenty feet by thirty feet wide, we knew they would need little space before spring. Tony built two wooden dens, each big enough for one bear. We filled them with straw, installed a water trough and again lined the fence with evergreen. One day, in late October, volunteers and the veterinarian gathered to move the bears.

The bears were much larger now, and fat. Their coats were full and gleaming with health. We knew they had been happy; we had watched them playing, eating, sleeping. Not, we were to learn, necessarily with complete understanding —

We decided that when we moved them, we would likely have to tranquilize them, at least partially. And that would be the time to take advantage of the situation and give Nungoon a vasectomy. After all, we did not need a natural increase of bears. And we had dutifully read and believed some biologists: a male bear has no part in bringing up his cubs. He will often kill them. The female is patient for only a couple of years. Anyway, we had read enough evidence that we did not want, or need, any bear cubs. Therefore, Ian White, our patient veterinarian, was ready to do the vasectomy.

Nungoon was given the tranquilizer, and then we waited for it to take effect. Bears resist mightily. We waited. Finally, he dozed off and the men dragged him out of the enclosure. Ian performed the vasectomy operation. No cubs.

While Nungoon was unconscious, we were able to look at him closely. Ian pulled the bear's lips back — and we saw — all his teeth were gone. The upper canines had not been pulled but cut with horn shears, leaving only stumps. The gums were slashed, damaged, hardened with use. The claws had been yanked out.

Nungoon had grown; six men were needed to lift him into the live-trap, put it on the back of the trailer and take him across the valley to the new enclosure. There, they laid him down in front of one of the dens, closed the gate, locked it and left.

Giisus had no doubt been aware of all the activity which had surrounded Nungoon. She was wary. Perhaps she thought something utterly dreadful had happened to him. For whatever reason, she was not going to be moved. She backed into the den and would not come out. Coaxing was useless. We had to leave her. For several days she sat, unmoving, not eating, back in the darkness. Several days. Two weeks.

Then, perhaps communicating with Nungoon in a way we humans cannot understand, she was suddenly willing. We had decided that we would have to force the situation — tranquilize her, carry her as we had carried Nungoon, unconscious, to the new enclosure. The volunteers and Ian came again. They opened the gate and pushed the cage against the opening. Very quietly, with immense dignity, she walked into it. The men closed the door, put the cage onto the trailer and once more made the trip across the valley. There, the trap open, she walked quietly into her new home.

Nungoon had been asleep. As she stood, perhaps uncertain, he oozed out of his den and came to her. For a moment their noses touched, and then, together, they squeezed into one of the dens.

The fact that the den was not supposed to be big enough for two bears bothered them not at all.

After the scarlet and gold of autumn, the whirling winds of November stripped the trees of their leaves and left them tall and dark on the hillsides. We wondered if the bears would hibernate. Was hibernation something which they had known in their years of captivity? The two dens were stuffed with straw, and waiting — one for each bear. They did the impossible and squeezed again into one. They slept.

Then quietly, softly, winter came. Great flakes of snow drifted from the low, grey clouds, settling gently on every branch and twig.

At first, as the cold came, there was a great quietness. Then the winter storms set in, filling the air with whirling whiteness, sweeping the deep snow into dunes, filling every crevice. Christmas Eve — and the singing of the wolves into the night-time.

Tony regularly inspected the bear enclosure. Once, he said, "Nungoon has shifted into his own den." We didn't think much about it. Both bears still slept.

January. The snow was deep and the cold, bitter. I was working close to the wood fire in the house. Stamping snow from his boots, Tony came in. He said: "I was just over with the bears. I think I heard a little whimper."

We rationalized: Giisus shifting in her sleep? Nothing more. Nungoon had had a vasectomy — hadn't he? Cubs are usually born, weighing under a pound, in the warmth of the hibernating den, the mother conscious enough to nurse it and keep it close to her body for the weeks of winter. One ancient native belief states that the cub is born formless and that, through those long first weeks, the mother gradually licks it into shape. Certainly, according to most biologists, the father has no part in the nurture of raising of cubs. He can even be a killer.

We speculated: could Giisus really have a cub? No matter how polite we might try to be to her, no three hundred pound mother bear is going to allow a mere human to push her aside to confirm the existence of a cub — but, by the end of the month, Tony heard enough whimpering that all doubt had vanished. Vasectomy too late, delayed implantation a fact of life with bears — a cub had been born.

We phoned the veterinarian. We didn't blame him — just phoned him! "We were too late with the vasectomy. Giisus has a cub."

He said, "Oops!"

So the cub acquired his name: Oops.

Once there was the story of two bears. Now it is the story of three.

January through February, the hibernation continued. Though the cub could be heard, he could not be seen. By March, our curiosity increased substantially. One of the men managed to rig up a camera (an old one, which would not, if it happened to be swatted by an angry mother bear, be a huge loss to the world of photography) and lowered it into the den. He managed to get one picture — an excellent picture — of a surprised looking very small cub with his mother nosing him aside.

By April, the hibernation was over and all three bears were awake and out. Oops was small, brown like his mother and convinced that the world was a wonderful place, he understood it completely, and no matter what biologists say, his father would never hurt him. And his father didn't. Patiently, he allowed the cub to climb over him, under him, attempt to share his food, as though they both thoroughly enjoyed each other.

The relationship precipitated the second name change. Though, in more formal occasions, Nungoon and Giisus have retained their native names, they have become known as Mama Bear and Papa Bear. For this demeaning change we have received criticism — even in a letter to the editor in one of our local papers. The claim was that we sentimentalized bears and so should not be allowed to give them sanctuary — obviously, we know nothing about bears. Be that as it may, we now have Mama and Papa bear.

The matter of building a large enclosure around the smaller one where the bears now lived became a matter of first concern. Tony and Janet worked at our sawmill, cutting posts, cutting boards. Rolls of fencing were brought by big trucks. The path that the fence would follow through the bush had to be cleared and dug out so that fencing could be put underground, as well as high above ground. The enclosure would be about four acres — woods, bush, a piece of meadow and a pond. Tony took the excavator and diverted the beaver creek enough to make a large pond, which he also dug. Volunteers helped, but the work was difficult — and kept being interrupted, because it was spring and baby animals were arriving at the Sanctuary, while some of the creatures we had wintered had to be taken long distances to be set free. We explained to the bears that, even though they were in a small place, they were not in a dark barn in Detroit, and the future was rather wonderful.

Oops continued to grow. Though the fence around their enclosure was certainly secure for the big bears, Oops could discover and squeeze through incredibly small gaps — which he did. He would be seen, outside the enclosure, up a small tree, or sitting on the roof, or scampering around the outside of the fence. His mother didn't seem to panic. Likely she knew he would come home for a meal! Which he always did.

All through the hot summer, when other animals were not demanding attention, the work on the enclosure went on. By September, we were getting anxious. I remember how it was: one day, when Tony and Janet were cutting the last joining poles for the fencing, a car came whipping up the lane. The driver was in a semi-panic. He had just hit a bear cub, out by the Rosseau cemetery. The cub had appeared suddenly, out of nowhere (they do) — he hit. It rolled under his car. At first he thought it was dead but it had come to, pulled itself to the edge of the road and tumbled down the ditch into the bulrushes — where it still was. Could we help?

Thus, horn blaring to indicate emergency, I drove down the primitive road to the bear enclosure construction site. Alerted, Tony and Janet hastened, of course, to the rescue. And, once again, their work on the enclosure was interrupted.

First, of course, they had to catch the cub. Because he was badly hurt, and they had a cage and a catch pole, that was accomplished relatively quickly. They rushed the cub to the Parry Sound Animal

Hospital. An X-ray showed a badly broken leg — however, the insertion of a plate would have to wait until the shock wore off. The cub (called Moses, because he had been pulled out of the bullrushes) was brought home. A small temporary enclosure had to be filled with hay. He was made comfortable, offered food and water and privacy. Tony and Janet still had an hour of the day left. So — back to the sawmill.

In the outdoor small enclosure Oops continued his mischievous ways. However, no matter how bratty Oops was being, he did manage to make one contribution to the world of medical knowledge: he donated blood.

A few days later we were called to Port Carling to pick up a tiny cub found lying beside the highway. Because of the proximity to a good deal of traffic, we presumed the cub had been hit by a car — until we found no sign of any wound or abrasion on his body. We also discovered that he would not have had the strength to walk into the path of anything, not even a pushcart. He was almost dead of starvation. Those few years ago, the Spring Bear Hunt was still permitted from the middle of April until the middle of June — that period of time when the young are utterly dependent on their mothers. And many, many mothers were shot (from shooting distance, differentiating between sexes is impossible) leaving orphaned cubs vulnerable to starvation. We have no way of knowing, of course, but the degree of starvation this little cub was enduring was the result of a much longer period of time than just the end of the fall hunt. We picked him up and carried him in our arms.

He had to have the lamb milk replacer dribbled through his blue lips. He had to be kept very, very warm. Day after day, night after night, Tony nursed him. No improvement. We took him to see our veterinarian, Ian White. He was puzzled so discussed the bear's condition with a veterinarian wild-animal specialist in Toronto. He suggested blood tests to see if something more than only starvation was threatening the cub's life. For comparison, he needed blood from a healthy cub. No cub on earth was more healthy than Oops.

Capturing a very lively cub, without incurring the wrath of the mother bear, is rather a challenge. We tried to lure Oops over the fence — a forbidden adventure he usually enjoyed. That day he stayed in the enclosure, the demure picture of the best, most obedient, little bear cub that had ever been given life. I don't really think he at all understood why the veterinarian was standing back, tranquilizing needle in hand, waiting. Up the inside of the fence and down, across the roof of the dens, sitting on the roof of the dens, going in to see his mother in one of the dens — he had a wonderful time. Then, finally, he made a mistake. He tried to squeeze between the back walls of the dens and the enclosure fence. And he stuck. Ian was able to give him the tranquilizer and Oops dropped off to sleep. Tony cut away a very small section of the wire and pulled him out. While Ian took the blood from Oops, Tony mended the fence. By the time Oops recovered, the fence was mended. Tony opened the gate properly and put him in. A little groggily, he sought the comfort of his mother.

The vial of healthy blood was taken, with the starving cub, down to the specialist in Toronto. Even with the best of care, all we could do was useless. The cub died. But Oops had done his best.

During all the interruptions, the construction of the enclosure continued. The posts were put in, the joining rails added, the fencing rolled into place ready to be erected. Not only did the fence have to be some twelve feet high, it had to be at least five feet flat on the ground, dug in and covered around the entire perimeter, so the bears could not dig out. And, once the fence was hung all around, large, four foot wide strips of smooth steel (donated), had to rim the tops — no climbing out. By the end of the summer, everything was ready. We all paced carefully around the outside of the enclosure, and around the inside. We tried to think like bears — was any escape possible? Not that we could detect.

TWO MIXED-BREED DOGS

Bruno — a coy-dog.

Silver — a wolf-dog.

Bill — a wolf-dog.

Brandy — a clever coyote.

WOLVES

Eclipse

Luna

The release of a fox.

Beaver at work.

Beavers at play.

Bear cub resting.

Mama & Papa bear with offspring.

Moose — one of Tony's favourite animals.

Coyote going free.

ON THE EDGE OF THE WILD

Bridge across a beaver dam.

The pond, some thirty feet by twenty feet, filled with water. Because of my obsession with beavers, when people admired the pond (then, and still), I explained how the water level was maintained by the beavers living in the tiny "lake" they had made in our valley. To this day, even in times of drought, the bear pond is full of water. Good beavers.

The day had come — the day we would open the door of the small enclosure and the bears could come out into the trees and the meadow, and the pond. For the great occasion, all the people who had helped with the construction gathered — some of us clutching cameras. We suspected that the opening of that gate would be as exciting to the bears as it was to us.

Going in, closing the gate to the big enclosure, Tony opened the gate to the small one, and stood back. Slowly, Papa Bear wandered to the opening — and stood — his nose to the ground, scenting.

Then, ponderously, he moved along the outside of the first fence, and, still very slowly, still with his nose to the ground, started into the darkness of the evergreens. Mama Bear followed, pausing longer at the gate, her head up for a moment, looking at us, at the space around her.

I wondered what they were thinking. Never before had they known space, acres of space. Never before had they been able to touch living trees, to feel the thick drifts of pine needles beneath their feet. Never before had they been able to wander into the sunshine and gaze on the grass in a meadow.

Like a child sharing an exciting adventure, Oops bounced everywhere.

They scented the pond. We moved down the hill to where it was — how would the bears react? React to something bigger than a bathtub?

Eagerly, joyously, Oops plunged into the water, swam in great circles; Mama stood at the edge.

She called him. He ignored her. She waded in to her knees. Next time he swam by, she grabbed him by the scruff of the neck and hauled him out. On the bank, he shook himself and plunged in again. This time, she followed, swimming in one circle and out again, shaking the water from her coat — pausing — and in again! The two brown bears swam around and around. Papa watched.

He watched — put one foot in and drew back. Then both feet — and inched forward. Rather like the launching of the Titanic, the big bear surged into the pond. And swam. For a long while, we watched and then we left them to their joy.

So, there they lived, the three bears in an ideal bear enclosure, as close to living in the wild as it is possible to be. But not free. And certainly not typical.

Giisus was the only one of the bears which reacted to the situation in anything like a normal fashion. She was not, and still is not, in any way tame or accepting of human companionship. She seemed to behave with her cub very like the way she would have behaved had she been alone with him in some wilderness place. While he was small, she suckled him. She allowed him to follow her everywhere. She let him feed in the meadow, where she still wandered out to eat the grass and roots. She shared her bowls of food with him. When he was disobedient in any way, she whacked him — hard! Like any cub, and like a human child, he whimpered and sulked. Unlike a bear cub, he sought consolation with his father. With Oops, Nungoon wrestled, gently. He allowed Oops to climb all over him. In the warm sunshine, Oops often slept on top of him.

That relationship was unnatural, and though we watched closely, we did not ever see Nungoon deviate from his tolerance. Nevertheless, we saw that Oops was pressing hard. In the wild, he would have been with his mother for perhaps sixteen to eighteen months, even a couple of years. Then she might drive him away. We had decisions to make — could we set Oops free?

Oops was growing. His colour changed from the red-brown which he had inherited from his mother, to the black which was his father's colour. His temperament was unpredictable. Had he been a human child, we would have classed him as a spoiled brat — perhaps sometimes a cute child, but having some dangerous potential as a big bear. He did not like people. In that, his mother had taught him well. But could he survive in the wild?

We thought that he could and would. Other creatures which, though raised in captivity, had been given their freedom, and we had been able to monitor their progress; the survival rate was high. Was it incorrect to assume that, because a skunk, admittedly not the brightest of all creatures, could easily make its own way in the wild, a bear, surely one of the most intelligent of all creatures, could not?

We consulted Lyn Rogers in Minnesota, one of the leading bear experts and researchers in North America. We explained the circumstances and asked the question. "Would he have a good chance, or should we keep him?"

The answer came back firmly. "You have no right to keep him. Let him go."

Our local government officials have a much more political answer, "You can't let him go", they stated severely. "He's an American bear."

We decided he had dual citizenship and began to make plans for his release.

Tony studied maps. We wanted a place where food was plentiful, bears were not, and humans were non-existent, especially when help from officials who would have some information was withheld. Tony searched, questioned, was careful. After long consideration, he put his finger on one place on the map. A good five hours drive from the Sanctuary.

The day came. Mama and Papa were confined in the small enclosure. Oops was tranquilized, and, while he slept, loaded into the large live-trap, firmly secured and hoisted into the back of the van.

From the barn doors where, those three years before we had watched the two mutilated bears arrive, we watched the van drive around the curve, and up, over the hilltop, and out of sight.

If Nungoon and Giisus missed Oops, they gave no sign of it. Once more allowed into the big enclosure, they proceeded with their feeding, roaming, sunning . . . quiet and seemingly very content.

All day long I was anxious, wondering what was happening. The work at the Sanctuary had to be done, the young had to be fed and cleaned, new residents welcomed in, people to meet and talk to — and all the time wondering about Oops. All day. Late afternoon. I knew that the return could not be until early evening. That, too, came and passed. The sun went down. Darkness came. Stars. Quiet. Almost midnight and the sound of the van coming over the hill . . .

When, at the end of a trail barely passable by the van, the trap had been lifted out and opened, Oops never hesitated. Not once looking back, he loped off into the deep green forest — free.

Later, as I listened to the valley quiet, and knew that somewhere far away, Oops was wandering in that same quietness, I was glad for him. And I thought about that dark stable where his parents had grown up, and the bright lights of the roaring city. Now, quiet.

Today, the snow is falling and the cold is intense. The two old bears are sound asleep in their dens. They have been asleep since early November and will likely stay asleep until late March.

We did our best for Oops. Through all the long summers and the long winters which stretch ahead, we will continue to do the best we can for Mama and Papa bears.

14 – Shaman Bear

The recorded message on the telephone was very succinct: "We're bringing you a bear."

And, several hours later, a car, pulling a large round live trap behind it, pulled into the lane of the Sanctuary. In the live trap was a very large, quiet, bear.

Now, my dogs and I have a great deal in common. They like to eat good food, sleep in a warm place and take long walks into the country. But we are also mixtures of breeding. The dogs are hound/ labrador/ collie/ whatever — mixtures. I am English, perhaps a touch of Scottish, and having had a great, great grandmother who, I am told, spoke with such an Irish accent that no-one could understand her. The surname bestowed on me is a French/Belgian mixture. Along with the beaver, I am Canadian. But after examining a family tree, which an uncle had given me, I found one twig which seemed to indicate that, away back in the dim distance, maybe and perhaps, a Viking — long, long ago.

I rather hope so. All those assorted ancestors, because they lived in the Northern hemisphere, likely held one belief about bears — certainly the Vikings did. This belief has been articulated in more recent centuries by some of our Native Canadians.

I'm not sure if my dogs care one way or the other — but one never knows.

Peoples of the Northern Hemisphere, all through Scandinavia, across Europe, across Russia, across the Bering Sea to North America — and even as far south as India and the Orient — these peoples looked upward and named two of the constellations moving around the night-time skies, the Great Bear (Ursa Major) and the Little Bear (Ursa Minor), identifying them in legends as the heart of all creation. The Bears.

Nor did that shared symbolism confine itself to the moving stars; the bear stood at the centre of the mystery of life itself. As, at the approach of the bleak, cold winter, the bear went into the darkness of the cave and slept — so, the stories said, is human death. As, at the approach of spring with the warmth of the sun, the bear awakens and returns to the wild — so, too, is human resurrection.

But just now, at this particular moment in history, a big, black bear was lying quietly in a round live trap, behind a car, in the lane of the Sanctuary. And, three humans stood beside it; two young, good looking native men, waiting for me to say something, and me, sensing that this was in some way a very special bear.

They told me that it was. "This is a special bear", one young man spoke at last, "and it must not die".

I assured them that we would give the bear the very best of care, and certainly not kill it. When it was well, it would go free. They listened and seemed to believe me. Even, perhaps, to trust me a little.

"His front leg is hurt." The young men were still watching me. "The people wanted to kill him." Still, unmoving, the bear, his eyes alert and intelligent, was watching me, too. Again, the emphatic statement: "He is a special bear. He must not die".

Then they told me his story. Limping badly, the bear had come into the settlement and begun to feed from the apple trees. Because of his obviously severe and painful injury, the people expected him to be unpredictable and therefore, dangerous.

"We have children around. What if he gets one of them? Shoot him!"

"No. He is Mishomis — Grandfather". While others laughed, the two men went in search of a live trap. People kept complaining to the local authorities. The Conservation Officer shrugged. "Why don't you just shoot him? No problem — it's bear season. Find the right market and sell his paws and gall bladder and make yourselves some money!?" Yes, the officer actually did say that!

However, a live trap finally rented and hauled into the settlement, and suitably baited with the biggest, juiciest apples, was totally ignored by the bear. For several days, still dragging the injured front leg, the bear continued to ignore it. Then the time ran out. Complaints reached the Ontario Provincial Police. Humans panicked. The police had to notify the young men.

"We're on our way out. We have to shoot the bear."

The settlement was less than half an hour from the police station. Quite matter of factly, the young men told me — "We went to the bear and told him what was going to happen. Within twenty minutes he went into the trap and lay down and went to sleep" — a pause — "we have named him Shaman".

Now, all bears are special — different from each other perhaps, but special. Over the long years I have raised and known literally dozens of bears — each, special. But not, I suppose, necessarily a shaman. According to the dictionary, a shaman is an individual who can communicate with humans, with spirits and with animals — an individual, sometimes in human form, sometimes in animal form, of great wisdom and power. Perhaps a shaman is somewhat like a saint — though I have never met a really saintly bear — nor a saintly human, for that matter. Anyway, as we began to move Shaman into the pen in which he would have to stay until he was well, he behaved much like a perfectly normal bear — handicapped only by the broken, dragging front leg.

Catherine, one of our young volunteers, had prepared a place for him — lots of big evergreen branches under which he could hide; deep, warm straw, and a large enough place that he would be able to move easily — but not so large that he would move swiftly and further damage his leg.

The only problem with the enclosure was its location — the trap had to be detached from the car, pulled, lifted, manipulated around two very tight corners, down a somewhat narrow hallway — and then, the trap opening had to be aligned perfectly with the pen opening. This was to be accomplished with an increasingly angry bear which, though lame, was highly incensed at the rough treatment and making no saintly secret of the fact. He huffed. With his good paw, he attempted to swat. He shifted his weight.

Eventually, he was in the enclosure. There, we could see his injury more clearly. The break was bad. His leg hung, useless and dragging. He rushed into the shadows of the evergreens, and lay — huffing and very angry. We gave him heaps of apples. That calmed him. He ate.

Over the days which followed, we watched him closely. The leg was useless. In the wild he would not be able to run well or dig or climb or swim. His future: forever in captivity? He hated captivity.

We telephoned our veterinarian clinic. On Thanksgiving day two of the veterinarians, Dr. White and Dr. Jones, both very compassionate men who have over the years given invaluable service to the animals in the Sanctuary, came to see Shaman. Backing the bear into a corner of his pen, they managed to get a tranquilizing needle through the wiring and into the bear; then, when we were absolutely certain he was utterly and completely asleep, we opened the enclosure door and dragged him out into the hallway. The doctors knelt beside him and began examining the leg. They were both very quiet.

"How," I asked, finally, breaking the long silence, "could he break and twist a leg so badly?"

"Probably caught in some kind of leg-hold trap", Dr. White said, "and thrashed around trying to tear himself out of the steel teeth." Severely injured, broken and blood soaked , he had managed. For about a month he had tried to live in the wild — then, desperately hungry, he had found the apple trees in the settlement.

Now he was with us. The veterinarians finally stood up, looked at each other and for a long moment, were silent. Then, "The break is not that recent — the bone between the wrist and the elbow has broken away and slipped up and begun to fuse with the bone between the elbow and shoulder. And his paw has begun to atrophy."

My heart sank. Now what?

The damage was so severe that only an extensive and extremely expensive operation could hope to mend it — if even that could. That operation would have to be performed by a wildlife specialist. We all remembered the time we had had a small bear cub, with far less extensive injuries, and when applying to the same experts, had been told: "It's not worth it. Have the cub 'put down'". The cub lived and had finally gone free. But we were hesitant to approach the same authorities — and, they were the only ones available — about this large, belligerent bear.

We returned him to his enclosure. Left him more apples. And decided we would think about it. What to do?

Because we were wintering half a dozen small orphaned cubs, and were short of space, we moved Shaman to an outdoor pen, where everything would be quiet and, well fed, he would likely hibernate. He had a dark den under evergreen branches. When winter swirled out of the north, he crawled into his den and slept.

Snow fell and fell. His den was buried. The temperature dropped. On clear nights it was often thirty below. No sound. No movement. He slept. All the weeks, all the months of winter — he slept.

Spring came. The snow melted. The wind was soft and warm. Shaman woke.

We opened the enclosure door.

On four strong legs, Shaman walked away into the wilderness.

15 – Nim-keas-quay

When I remember the bear cub, Nim-keas-quay, I somehow think of a deep darkness and a long, slow beginning. That night, when she came to the Sanctuary, the darkness was very literal, and — I cannot explain why — even vaguely frightening.

Toward evening, early in April, the phone rang. A masculine voice, a bit abrupt, brusque, said:

"We're working up the highway by Huntsville. Found a cub. Almost dead, I think. She was layin' in the grass. Got her here. Do you want her?"

Yes. Cubs are born in January. That year no tiny cubs had come to us. Though I would never wish harm to a mother bear, or her cubs, I had rather missed having a small, feisty, little cub. As a rather severe friend said later, "You may have mentioned to God that you wanted to look after a bear cub. He told you to wait until one needed you." Okay. Point taken. But now this cub needed help.

"How big is she?"

"Don't know — twenty-five, thirty pounds maybe".

Too large to be a cub from this year, she had to be a starved yearling. Perhaps her mother had been killed in the spring bear hunt, or the fall — perhaps her mother had been killed on the highway. Those facts we could not know. We did know that, had she had a mother she would have weighed between sixty or seventy pounds, and she would not be lying weakly beside an area of noisy road construction.

"Where are you?"

He gave directions — north of Huntsville on the 60 highway to Novar, a left turn on Old Boundary Road (gravel) for about ten miles to a blue garage, where the men would be waiting for me and would take me the rest of the way. A long trip. The early darkness was already beginning to close in.

I am blessed with more cousins than I can really count — first, second and third cousins and a few honourary cousins who are their friends and mine. They all live a good distance from Muskoka. However, as I put down the phone, picked up my jacket and stepped out onto the porch, a pickup truck came down the driveway. They had come all the way from Hamilton. I gave them a moment to stretch after their long drive and then announced:

"Get a strong, sturdy carrying cage from the barn. I'll get blankets. We're going to pick up a bear cub".

Randy and Don were young men in their early twenties, young enough to enjoy the adventure, strong enough if I needed help with the cub, and resilient enough to laugh, get the cage and climb back into their truck for another trip.

As we headed out through the swiftly gathering dusk, up the winding road toward Huntsville, I explained to them where the bear had been found, the probable reason why it was so small and so sick. I gave Don the map I had drawn, told them where the men would meet us and settled back to wonder — having help in a pickup was very, very nice.

Long before we reached Novar, the darkness had closed in. We overshot the turn to Old Boundary Road, had to turn around and come back. For the first couple of miles we passed small houses but after that only miles of winding dirt road, woods and brush. Not even stars in the sky. The only light was the long beam of the headlights, cutting ahead of us, showing the shadows of the trees flicking by and the winding of the narrow road. Ten miles is not really far, but in the loneliness and the darkness, the minutes began to seem like hours. Finally, on the right hand side, a single light bulb shone above a garage door and a blue truck.

We pulled in. Randy rolled down the window. One of the men came over. "Follow us," he said, "it's a bit further."

Randy backed out, waited for the men to pile into the blue truck and, when it was on the road, we began to follow it.

I know that the night could not have become much darker, but somehow it seemed to. The road became narrower, brush pushing in so close that it whisked against the truck doors. We didn't talk anymore. Just followed. And followed. Through the darkness. On and on. No houses. No camps. Just two red tail lights we were following.

And then — around a sudden curve, again, lights — a house with lights shining dimly through the windows, and a porch, where again, a single light bulb shone. We followed down a lane and stopped by the porch.

One of the men came back. "She's out in the shed. Come on, follow us."

We did — behind the house, so that once more the only light was the beam of the flashlight, past a couple of small barns, and into a shed. In a small, straw-filled bin, so weak she could not lift her head, lay a small bear cub.

So small and so weak. Though her eyes stared into the light from the flashlight, she did not move. I was conscious of the great darkness of the shed, of the woods, of all the world around us, stretching to the very stars. Out of that great darkness had come one sick bear cub — who needed us.

Randy and Don were talking quietly to the men. I knelt to stroke the cub. She made no response. "Let's get her home quickly," I said, whispering, though I don't know why. Randy and Don lifted her. I wrapped her in a warm blanket. They put her in the cage which we wrapped again, and walked out of the shed, back by the dark barns, to the truck.

The room in my house which I self-consciously called my "studio" was the only room which was warm and also offered complete privacy. No very sick cub would want to contend with several curious dogs and a beaver who was certain he controlled admission to the house. So we carried the cage into the studio, opened its door, and made a more careful examination of the cub. She simply lay still. I offered her food. She did not even sniff it. Sweet food with honey. Generally a bear will make an effort for honey! No response — and so we wrapped her warmly and left her to the quiet, warm, darkness. We did not even close the door.

Though I slept on just the other side of the door, I did not hear her stir in the night.

Just as dawn began to come, I went into the studio. She lifted her head, and slowly, with utter weakness, rose to her feet. Slowly, so slowly, she crawled under the typewriter desk and into the little compartment there — not once taking her eyes off me, wanting to somehow be away from me, but very weak. However, since the light was coming, I saw her clearly for the first time. She was thin, so thin that ribs showed beneath her sparse, grungy coat, brown-black and dry. Her face was thin, shrunken like that on an old, old beggar; her eyes were dull. Her gums were bluish-white. When the sun was high in the sky and the air was warm, we carried her to a small hay-filled pen outside, covered her with evergreens and blankets. We carried her, in our arms — a wild bear from the dark winter wilderness.

Then, for one hour, and then two, we dribbled goat's milk down her throat.

I phoned our veterinarian. After I had described her condition as carefully as I could, he paused. Then he said, quite gently, "I know you like bears but don't get too attached to this one. She sounds too sick. I doubt that she will live."

Both the veterinarian and I like it when, in a situation such as this, he is wrong.

We named her Nim-keas-quay: Ojibway for Little Strong Woman. She would have to be strong to fight her way back from the very edge of death, to live once again out in the freedom of the forest. We humans owed her that life; if her mother had been killed by a car, or if a brave hunter had paid $20.25 for the permission legally to kill her mother — we still owed her.

For several days she stayed hidden in the straw, the blankets and her substitute evergreen woods.

When we had to excavate for her and hold her while we forced goat's milk and Pablum down her throat, she made no objections. Every time we approached the cage, bowl of food in hand, we hoped she would fight us. When she raised her head in anticipation, we felt we were winning. When she managed to huff a little, and then try to shuffle backwards, we would always leave a bowl full of food in with her. Finally, she stood tottering on her feet. The morning we came out and found the bowl of food licked clean, we felt we were going to win.

That small victory was followed by days and days of patient nursing, of slow, slow recovery. Randy and Don had to return to Hamilton; other humans stopped by to help. Nim-keas-quay stayed hidden in her log and the evergreen, coming out only to lick her bowl of Pablum. As is so much of working with wild creatures: no drama — just day after day after day a slow, sometimes almost not noticeable improvement — but watching for the small signs.

I could tell by the disturbance in the hay and the evergreens that she was moving around at night when no human was there to watch her. So we prepared a larger enclosure where she could walk freely, even paddle in a small pond, or hide when she wanted to in a small shed full of hay and evergreen. Not easily carried anymore, friends helped me to lure her into a small cage so we could move her to the new place. Released there, she scampered into the evergreen. And, like a good bear, she hid.

We fed her apples and kibble and she ate hungrily. She ate berries and other fruit. She ate as much as we could give her. Of course, she began to put on weight. Her ribs disappeared. One day, when the sun was out, we noticed that her coat was beginning to shine. The proper pink returned to her gums. Her face filled out.

Her eyes. The mood of a bear is often betrayed in the eyes — sometimes speculative, sometimes a sideways look which means: "Be careful! I'm meditating danger!", the dull look of sickness. Nim-keas-quay's eyes began to sparkle a little, watching us carefully, calculating every move we made. When I cleaned the pen (and she was becoming very excellent at creating scat!), I had to keep an eye on her always — she would huff warnings if I moved too quickly, or came too close.

The day she made a run at me, I knew she would once again be a wild bear.

For most of our wild creatures, release back to the woods comes at about the same time they would be naturally leaving their mothers; that is a time when, if they were raised wild, they want independence, and they are able to care for themselves. For bear cubs, the time is somewhere between sixteen and eighteen months. We also have to take into consideration the health of the cub — and, at that time — the date of the end of the spring bear hunt. That was about the middle of June. Though Nim-keas-quay was showing every evidence of being able to look after herself, she had to wait.

That year, by the middle of June, there were lots of berries, green things, roots and buds in the woods. All our experience with cubs assured us that she would do well. She was no longer approachable. I was very glad of the help of two young men — not cousins this time — who came to help move her from the pen into a cage which we could transport in the truck.

She had no inclination to be cooperative. The open cage, baited with her favourite berries, was put into her enclosure. She huffed at it and went into her shed. With long poles, we tried to ease her toward it. With complete contempt she pushed them aside. Finally, the men built a small chute from the pen door into the cage. She didn't resist the temptation of the open door — down the chute and into the cage. We closed the door and locked it.

Angrily, she huffed.

It was late afternoon. In the back of the truck, the cage was carefully covered — though, from time to time, it rocked with the resentment of one angry bear. She had to go back to the bush, far north-east of Huntsville, where she would not encounter humans, or traffic. And so, as the sun began to sink lower in the sky, the truck finally turned onto a long, narrow dirt road which wound up to the north of Algonquin Park, then turned between high hills where no humans lived.

Trees, wetlands and the quiet of the wild.

Finally, where a small lake bordered a grassy hillside, the cage was taken out of the truck, uncovered and turned toward the hillside. The door was opened.

Nim-keas-quay was free. She ran up the hillside — up toward the thickening darkness of the woods. At the top she stopped, for a brief moment, and looked back down the hill. Then, she was gone.

As the truck returned down the narrow, winding road, darkness closed in.

16 – The Story of Fuzz, the Bear

Cold — the thick ice on the lakes heaved, the sound of its breaking almost shattering the stars. Trees snapped and split. Evergreen branches, snow packed, hung motionless. The white hills were still beneath the clear, full moon — cold, January cold. Deep in a cave beneath the roots of a gigantic pine, a mother bear, sleeping, curled around the body of her tiny cub, new born, and warm.

Together they slept through the January cold. As he suckled her rich milk, she barely stirred. Then, while the February blizzards whirled from low grey clouds, they waited. The cub grew; in the darkness of the cave, she was aware of him, and of the great wilderness outside.

March, and a warm south wind. The cub opened his eyes to the tremendous bulk of his mother and began to hear the murmur of melting snow. And soon the cub was large enough to follow mother out of their winter shelter. After that there is little of his life we know for certain. He followed her. He continued to suckle her rich, warm milk. He slept close to the comforting heat of her body. He watched her as she grubbed for roots and growing green things. His small nose began to learn the smell of food.

Then, suddenly, the little bear was alone — utterly alone. We can know that he was not with his mother very long because, weeks and weeks later, he was still a very small bear. We cannot know why she suddenly disappeared. Bears have no real predators, except humans. For two and a half months, in the lovely springtime, bears are hunted. It is most likely that, during this period, some sports person bravely shot "his bear" and boasted about his prowess all summer long. But the little bear was alone and very small.

About his summer we may know nothing. He must have found some food because he did live.

Then it was September, and once more the sports people take to the woods. And all we may know is that there was another trauma. Bleeding and almost dead, the little bear was lying amidst the garbage at the Still River dump. There was a long jagged wound down his side. It was crawling with maggots.

Two Ontario Provincial Police officers found the little bear and saw he was not dead. Very thoroughly, before approaching the cub, they searched for the mother. Then, not finding her, they wrapped him in an old blanket and put him in the cruiser. Perhaps the rest was rather fun; turning on the siren, they sped down Highway 69 to the Parry Sound Animal Hospital.

Labour day was over. Most of the tourists have returned to the city. Quietness settles over Muskoka. The phone rings — and our diary begins to tell the story.

Tuesday, September 5. "Ontario Provincial Police, Still River, calling. Could you handle a baby bear when it recovers at the veterinarian's? " Yes, we could.

I visited the bear at Dr. White's office. It was lying in a metal cage, on a yellow towel, on its side. When I touched it, it made no move beyond opening its eyes. Dr. White said the wound was likely caused by an arrow. Since the bear was lying on that side, I couldn't see the wound there, but there was a sizable wound on the shoulder.

I went to the grocery store and bought Pablum, apple sauce and blueberries and took them back to the veterinarian's office. The bear made no attempt to sit up.

Friday, September 8. I phoned Dr. White's office to see how the bear was doing. "He has made some attempt to roll to his chest and is eating well."

Saturday, September 9. About 9:45. Dr. White and Laura-Lee brought the bear to the Sanctuary. He was still so passive he could be carried in a blanket. Around his neck was a little red collar. Downstairs in the barn is our "warm room" — a small room, with no windows to let in drafts, no other creatures near, and a heat lamp to keep anything comfortably warm no matter what the outside temperature might be. For the first time I saw the wound the arrow had made — an eight inch tear down his side, that had taken more than twenty stitches to close. Dr. White — Ian — says we will have to take the stitches out in about a week. Now the bear is lying on his side but is able to lift his head to lap food from his bowl. He doesn't object to having a towel tucked under one side to prop him up.

In honour of his O.P.P rescuers, we have decided to call him Fuzz.

Sunday, September 10. This morning when I went to see him, he hid his face in his paws. He is eating well, three bowls of food a day, as well as peaches. He huffs if I move too fast — that indicates, I think, that his energy and natural instincts are returning. He raised his head completely, very alert. At night he is covered with a blanket.

Monday, September 11. This morning the blanket was kicked off — so he had been moving in the night. He does not mind being propped up on both sides now so that he is resting on his chest and stomach. His head is quite under control — he holds it strongly. He shifts his back legs as though he is trying to stand. His front paws are moving — he uses them to steady his bowl. He made huffing noises and snapped his jaws at me. We were able to talk to him and rub him all over. We took off the red collar — it seemed wise to do it while we could still handle the bear. Fuzz ate all of his Pablum, plus peaches, bananas and a piece of apple.

Tuesday, September 12. When I went up this morning I found that Fuzz had moved himself from his corner and was lying with his nose against the door. I moved him back to the centre of the room and he did not resist.

I phoned the Still River Ontario Provincial Police to get the exact story about the discovery of the little bear. They reported that they had received a call that there was a wounded bear in the dump near Pickerel River. So they had driven out to investigate. They had made certain that no mother bear was in evidence, then approached the cub and picked it up. He was completely docile. The bear was taken straight to Parry Sound where Dr. Alan Christie (the retired veterinarian who still worked with Ian) did immediate surgery. The wound was several days old and full of maggots.

Today, Fuzz is much spunkier. When I tried to move him, he both huffed and snapped.

Wednesday, September 13. Fuzz was calm and friendlier this morning. He was lying on his side when I went in, but stretched his back legs straight out. He let me roll him on his stomach and prop him up. He ate well. His present diet is Rice Pablum, apple sauce, raw egg, fruit (blueberries, peaches, bananas), and dog kibble. Last night I left him a sticky bun for a treat but he did not eat it. He snapped and huffed at a staff member.

4 p.m. He shifted all the way across the room and was lying with his back legs out behind and his nose at the door. I lifted him back to the middle of the room and he made no protest. I sat with him as he ate and he seemed quite relaxed. He started to shift his back legs as though he were making little attempts to stand.

Thursday, September 14. Fuzz is moving all his legs. We decided to move him up to an outdoor pen. We filled the little shed with hay and put a few evergreen branches in the run. Then we lifted Fuzz in a sling and carried him down. He was curled up in the hay but he is trying to use his legs.

Friday, September 15. I pushed on Fuzz's back legs — they have a strong thrust. He tried to swat me with a front paw. He spent most of the day lying on his back with his round tummy exposed. I didn't turn him over for his evening meal. I wanted to see him do it himself.

Saturday, September 16. He could and did. This afternoon he "swam" out of the shed to the open pen, using all four legs to propel himself along. And, he was very feisty — he could move his head quickly and right around. We rolled him onto a blanket and moved him back into the shed for the night.

Monday, September 18. Fuzz moved himself all the way out of his shed, nearly the length of the enclosure and then back again.

Wednesday, September 20: Fuzz has been lying on his back "cycling" with his back feet.

Thursday, September 21. Ian came down to remove the stitches. We carried Fuzz out to the pen in a blanket. I held his head down under a towel. Laura Lee held his back feet. Ian worked. Fuzz bawled. Afterward, Fuzz rolled over on his stomach and scrambled, all four legs working, into his den. He has become much more active, moving around the den. Ian feels there is great improvement.

Friday, September 22. Fuzz is almost standing in the corner of his pen.

The improvement continued slowly. The diary entry for **Thursday, October 12** sounded a note of triumph:

> Saw him definitely walking. I have been wondering about the way his front paws toe in — but he was definitely and absolutely walking. Now I am leaving his food out in the cage so that he has to walk for it.

This was to become known as our first "Winter of the Bears." As Fuzz began to walk, he needed a larger area, so we built a larger pen in the barn. It was just as well — we would shortly be needing all the pens we had. We were wintering raccoons, four beavers, two coyotes, two Eastern Timber wolves, two Northern Timber wolves, a couple of fawns, skunks, squirrels and a dozen hawks, owls and ravens. Nevertheless, when we found another message on the answering machine, a second pen was begun beside the one for Fuzz. The message simply said: "We have a cub and are on our way over. You be home!"

This cub, again orphaned (the fall bear hunt was in progress), had wandered into a dumpster at the Grandview Estates in Huntsville, looking for food. Once in the dumpster, she could not get out. There she had remained for several days until someone called the Ministry of Natural Resources. A Conservation Officer was to come to dispose of her. Since that would likely mean shooting her, a gentleman who heard of her plight moved quickly, secured a cage from a nearby veterinarian, blithely passed himself off to waiting viewers as "the man who was to come for the bear", trapped her, put her in his van, and then phoned us.

When she arrived at the Sanctuary, she had to wait in the small cage for a few hours while the construction of the second bear pen was being completed. Then, after we had filled the manger with hay, we released her into it. Because that was the evening of the earthquake in San Francisco, we named her Quaker. She was very wary.

Again, less than a week had gone by before, returning home about nine o'clock in the evening, I found another message on my answering machine. This was a message from a Conservation Officer — one who cared. A sow with triplet cubs had been killed a couple of weeks ago. Two of the cubs had vanished into the bush (not likely to survive), but one with an injured foot had been found. It was now at the Centennial Animal Hospital in Bracebridge. When it was ready, would we be able to care for it? Yes.

So, by the 25th of October, when I picked up this little bear, whom the hospital staff had named, Yuri, we had three cubs in the barn.

Fuzz had improved so much that he had to be tranquilized to move him to his new pen. But now he had to walk the full length of the enclosure to get his food.

So began that winter of the bears. For about six weeks, every three days, the bandages on Yuri's foot had to be changed. This meant that I would get the salve and the bandages ready, then while someone, using a thick blanket held Yuri firmly pinned to the ground, I would cut away the old bandage, smear the salve onto the wound and then rebind it. I would be safely out of the pen before the very angry bear was released. Before long Yuri had removed the barrier between himself and Quaker and the two of them slept in the warm hay in the manger.

As the winter deepened, Fuzz, now walking very well, built himself a huge nest of hay at the back of the pen. He curled up on top of it. Gradually, he sank down into the middle so that he spent the winter sleeping like an old Canada Goose on her nest. And, though Quaker and Yuri showed no inclination to hibernate, Fuzz slept deeply, waking only if the weather happened to turn mild.. Even then, he ate very little food.

January — bitter cold. This is the time when mother bears, deeply asleep in their dens, give birth to their young. About eight ounces, and blind at birth, the tiny cubs are warm and cared for. A year ago, our cubs had been born with every right to live long and placid lives in the wilderness. Now, a year later, with their mothers murdered, they were captives.

Cold — and winter hunger. There was one more cub. Too hungry to hibernate, certainly not enough fat to keep him through the winter, not able to hunt for himself, this little bear discovered a deer yard where people put out grain for the deer.

He was a clever little bear. While grain wasn't his favourite food, he seemed to know it would keep him alive. The deer, however, were not enthusiastic about sharing food with a bear, however small. Consequently, he was live-trapped and brought to the Sanctuary. He was very small and, perhaps because he had been out in all the winter weather, his fur had grown very, very long. Nor did he take well to captivity. Though he enjoyed the food offered to him, he resented the caging. He paced, not slowly, but at a small determined gallop — up and down and up and down. Because we had no other enclosure, he was in the outdoor pen we had made earlier for Fuzz. We could see his restlessness and his long shaggy hair tossing about — his name was predictable: he was named Woodstock.

Once Yuri's leg was healed, the three cubs required nothing but feeding and cleaning. Fuzz was pretty much asleep. All we had to do was to wait for spring and plan their release. Looking for an area where the cubs could be released with some degree of safety, maps were spread out and research papers were consulted.. The place must be remote from hunters, have few other bears and offer a plentiful food supply. Such places are difficult to find.

March, and the scent of spring was in the air. The little bears began to grow restless. Yuri, unable to wait, tore a hole in the enclosure and set himself free. As far as we know, he still lives in the woods at the back of the Sanctuary, and, very welcome he is. Quaker, who was sharing his pen, had not left with him but grew even more cautious of us. Woodstock continued his wild pacing.

Fuzz awakened — but he couldn't walk. The slight inward curve which he had in his paws in the fall had, without exercise, grown much worse. Now he stumbled and blundered his way around in his pen. Obviously, he was in no condition to be set free.

About this time, a young woman who had been studying to be an animal technician, came to spend a little time working with the animals at the Sanctuary. She took a keen interest in the little bears, and particularly in Fuzz. Lisa kept notes on her experience at the Sanctuary:

> **Friday, March 30.** Last night I spoke to a veterinarian at Guelph University about Fuzz. She said there is a very expensive operation that can be done but its success depends upon follow-up procedures. Since Fuzz cannot be handled, therapy would be impossible so the alternative is euthanasia. We made a video tape of Fuzz walking on his wrists today and, watching it carefully, decided we would try to rehabilitate him through therapy. We will build him a large pen with some sort of pond or tub for him to swim in, and a swinging tire. We also thought that, by hiding food inside a hollow log or putting treats up high on a branch, this would force him to use his front paws. We hope that, by inducing Fuzz to use his paws, we can reverse the damage.

I, too, was keeping my usual journal:

> **Saturday, March 31.** We called Elva Wood, who has been practicing physiotherapy at the Parry Sound hospital for many years. We asked her if she would come and look at Fuzz and tell us what we might do for him. I'm going to pick her up in Parry Sound tomorrow afternoon.
>
> Elva is well know in the Parry Sound area for her skill in physiotherapy. There are literally hundreds of humans who owe her a great debt of gratitude for her skills, but this was the first time she had been confronted with a bear. She stood quietly, watching Fuzz move, stumble, fall over on his nose. Somewhat tensely, we watched her, waiting for her verdict.
>
> "I think," she said finally, "as long as those tendons are not fused, we can help him." At the very least, hope.

Lisa continued her notes:

Sunday, April 1. What a day! Judith Brocklehurst had an article in the *Toronto Star* about the Sanctuary today, saying we needed five thousand dollars to fly our bears, wolves and coyotes to their release sites. At noon a couple from Bracebridge drove up, took a tour of the place and wrote Audrey a cheque for five thousand dollars. Wow! There were also calls from several other people saying they were sending us cheques. Any money not used for this release we'll leave in the bank for Fuzz's release later, because when Elva looked at him today she said she doesn't see any reason we can't expect to rehabilitate him fully with splints. She seems to know exactly what she's going to do and how she's going to do it and even has the necessary materials to make the splints. What she wants to do is have a rigid section immobilizing his wrists in the proper position, with a flexible position above and below which will add stability while allowing movement and forcing him to use his leg, paw and claws.

We have been and will continue to document the whole process on video. Until she can feel his paws and see how much she can move them, Elva has no idea how long the whole rehabilitation will take. She feels his left paw is worse than his right. She also believes this was likely due to nerve damage from his wound, or possibly the condition is congenital. She doesn't think it would be this severe simply from lack of use. She also told us to expect Fuzz to be in some pain for awhile until he starts to adjust to the new position of his feet. Let's hope this works. It would be awful to put him through all this for nothing. He's been through so much already. Elva seems so confident though, I really believe she can do it.

Thursday, April 15. Audrey phoned Ian and told him about our plans for Fuzz. He was very enthusiastic. He will pick up Elva and bring her down to put the splints on the bear.

Quaker and Woodstock were flown north and released to begin again their lives in the wild. Letting them go free is an act of faith: we have done for them all that can be done; now we must commit them to the wilderness where they belong. We must trust their inherent intelligence, the knowledge they inherit from generations before them and we must trust the great wild land that we have chosen for them. We must trust the love of the Creator who made them.

Fuzz, alone now in the barn, still stumbled about the bear pen. While the other cubs ranged the forests, he faced a long summer of therapy. Having rejected the death sentence, it was now our responsibility to do everything we possibly could to prepare him for free life. He must be forced to use his paws.

Ian and Elva did their part splendidly. Every single week, all summer long, they came to the bear pen. Ian devised a needle on the end of a long pole so that he could inject the tranquilizer in the bear's rump from a safe distance. Once Fuzz was asleep, Elva went to work manipulating those front paws, moving them a bit more forward each week. When she was ready, Ian bound the paws, pulling and holding the tendons straight. When Fuzz awakened, he would blunder to his feet. And, as though he knew their importance (I think perhaps he did), and though he was perfectly capable of doing so, he never once attempted to remove the bindings.

Meanwhile, in order to make a free life eventually possible, we had to make his present life more difficult. Walking on smooth, hay-covered floor would have been easier; having to walk through strewn branches and over logs, some large and some small, would force him to use his paws. We covered the floor with various sizes of branches, logs, evergreen boughs. Fuzz had to work for his food, too. The apples and pears, peaches and berries — and sticky buns, for treats — were to be found only at the bottom of a tower of five or six rubber tires. Theoretically, he would have to climb the tires to get the food; actually, he usually scattered them. Either way, his feet were exercised. Finally, between his sleeping quarters and the outdoor pen, we built a swimming pool. To get outside — which he wanted badly to do — he had to swim.

In the pool we put a big bright, red ball, too big to bite but perfect for batting around. Though we never actually saw him playing with it, we know that during the night, he must have done so. Several times the ball was torn apart and had to be replaced.

And so, June, July and August passed. Then, one day, in early September, when Ian and Elva took the bindings off his feet, we all stood and watched Fuzz walking. No stumbling. No hesitation. Walking.

Ian said, "Time for him to go."

Autumn — frost was beginning to touch the tips of the maples with crimson. Mornings were sharp and clear. At last, Fuzz could return to the great waiting wilderness.

The release plans were completed. The day was set. As we met by the pen, Fuzz eyed us with some speculation. He had come to know very well what happened when he allowed that long pole with the needle on the end to come too near to him. Very deliberately, he retreated to a concrete corner of the pen and sat down. His rump was completely protected. He waited for our next move. Ian stationed himself and his pole at the strategic place. From the opposite side, cautiously, a Sanctuary worker approached the bear — carefully, very carefully. Fuzz weighed over one hundred pounds now, and had never been tamed. Closer. Fuzz decided it was time to stand up and make threatening noises. Ian jabbed him. Fuzz went to sleep.

Quickly, he was bundled into a cage and lifted into the back of the truck. I saw the black bundle that was Fuzz asleep, and then the doors of the truck were closed. He had lived with us for an entire year. Now, I would not see him again. The truck drove down the laneway, turned left, drove up and over the top of the hill and was gone. When a bear leaves, a strange loneliness falls over the Sanctuary, a loneliness containing both satisfaction and loss — and hope.

Practically, there was a bear pen to be cleaned!

Because Fuzz had been crippled, because he had overcome considerable odds, we wanted him released where he could be watched. Fortunately, good friends of the Sanctuary had moved to a very remote area in Northern Ontario. For Fuzz, they had located a ravine with a good cave for hibernation and lots of food — and no other bears. There, they could monitor Fuzz with field glasses.

The plan was excellent. During the trip North, Fuzz awakened and expressed his indignation by rocking, not merely his cage, but the entire back of the truck. Once they had arrived, the cage was carried down into the ravine to the foot of a steep hillside. We wanted a good video of Fuzz walking. We wanted to prove to all those who had said he should be euthanized that, yes, it had required time and ingenuity and the work and devotion of many people, but Fuzz made it. He walked.

The cage was carefully positioned so that, once the door was opened, Fuzz would choose to run in front of the camera. Otherwise he would have to climb the steep hillside. He climbed the hillside!

Winter came. Deep snow. Fierce cold. No sign of Fuzz. Had he stumbled on that steep hillside and fallen? Was he hibernating? All winter? All we knew was the great silence of the snowy hills. Spring. Melting snow. And, at last, tracks — the right front paw still very slightly crooked. Fuzz. Alive in the wilderness. Free.

17 – An Overbearing Situation
The AUTUMN and WINTER of 2001–2002

One would think that after thirty years in the business of rehabilitation I would have more common sense than to entitle my final newspaper column for *The Muskokan*, in their October edition, "The End of a Long and Busy Season". Even the temporizing in the first paragraph indicates a degree of naivety:

> Of course, 'The End' does not mean we are really at the end of anything much — just that, except for squirrels, rabbits and mice (who do not read biology books) we are not likely to have any more babies arrive. We will not be giving bottles, or getting up in the middle of the night for extra feedings, or desperately ordering more Esbilac because we have run out and the phone has just rung to inform us that a dozen more raccoon kits are on the way here to the Sanctuary. We will continue to clean cages, pick up food, tend the animals who must winter here, look after injured animals who arrive — and wait for spring.

The circumstances of the summer should have been a warning that no quiet time was likely to come. All during June, July, August and into September, this area suffered drought. Water levels fell (except where humans had been clever enough to allow the beavers to dam water), trees turned brown, the grass looked burned — and, out in the bush, no berries ripened. Prolonged drought — and that meant no food for the bears. In desperation, the bears sought food in the only places it could be found — in the garbage cottagers left around, bird feeders and barbeques.

This was the year the municipalities decided that all the dumps would be containerized, so even that source for food was no longer available to the bears. Extensive road construction, with constant dynamiting and destruction of habitat caused the already hungry bears to be disoriented.

The bears, hunting for the scraps which humans leave, came close to cottages and they even wandered into towns. Five bears were shot at the feed store in Parry Sound. According to the *Toronto Star* newspaper, fifteen bears were shot in Bancroft and over two hundred bears were shot between Sudbury and Timmins. Reports of bear shootings came in from across the province. People, uneducated in the ways of bears, panicked. And the bears died. Leaving orphaned cubs.

We should have anticipated something! However, our bear season began quietly enough in June. Sometime in early May, near Port Sydney, a mother bear was killed on the highway. Shortly afterward, a very tiny cub was seen in a Port Sydney neighbourhood, starving, lost and, without some help, doomed. Seeing a little cub, and wanting to help it, are only the beginnings of the long process of actually catching it. Even cubs can be very wary of humans. However, the people who live along River Valley Road, saw the cub and wanted to help. Their first call was to the Sanctuary.

Tony and Janet took a live trap over to River Valley Road, showed Jim and Carol (the people who had called) how to bait it and set it, and the wait began. One week. The cub was cautious. They kept the bait fresh. Two weeks. Occasionally they saw, or heard the cub. Three weeks. They varied the bait: always honey, but over bread or buns or peanut butter — anything that might tempt a bear. The location of the trap was changed. Once, they decided that the cub had disappeared — and then it was back. Four weeks. Unless it was caught soon, it would die.

And then, the telephone call: "He's in the trap!"

So the first bear of the year came to a small enclosure here. He would peer out of his hollow log, huff through extended lips, stomp his feet — in short, give his best imitation of a huge grizzly. He would retreat into his log and charge out again — huffing mightily. His imitation was actually quite impressive! Except the little black bear cub weighed only ten pounds.

Though we were sorry he had lost his parent, we thought that raising a single bear would be quite fun. In previous years, before the cancellation of the spring bear hunt, many small, orphaned cubs would be brought to us. Since the cancellation of the spring hunt, we had had only one cub — from Quebec, where the hunt continued. The Port Sydney cub had been orphaned by a road accident. We knew that he must not be tamed, that he, if he was going to survive when he was finally given his freedom, would have to be very wary of humans. Still we watched him, enjoyed him, and Tony fed him.

Had we known the devastation occurring amongst the bears here in Ontario, would we have agreed to help the cubs from Manitoba? Probably. An orphaned cub is an orphaned cub. Without help, it will die. Manitoba has no place for orphaned cubs.

Last year, when a telephone call asked if we would consider raising two orphaned baby otters, I did not hesitate. Of course we would take them! When I discovered that the call about the otters was from Creighton, Saskatchewan, I had to go hunting for my old school atlas to find exactly where Creighton was: northern Saskatchewan, right across the Manitoba border from Flin Flon. So, when another call came, this time from The Pas, Manitoba, I again had to consult the *Atlas for Canadian High Schools.*. The atlas is rather ancient and not many places in northern Manitoba were marked — but I did find The Pas, near the north end of Lake Winnipeg. We were to find that the flight from The Pas to Winnipeg took nine hours.

The Conservation Officer at The Pas was looking for a place to care for two orphaned cubs. Their mother had been shot — the spring bear hunt persists in Manitoba. The officer and Tony began searching for a way to transport two very small cubs, and, once again, as they had last year for the otters, Bear Skin Airlines came through. The cubs were flown to Winnipeg and then to North Bay. Late one night, three days after the first call, Tony met the plane in North Bay. The cubs arrived, both in a small blue transport barrel, easily carried by two people. (Will the return trip be so easy?)

The Manitoba cubs were very small, wiry individuals, dark, dark brown — one with a large white "V" on its chest, the other with a smaller "V". They were very vocal about their hunger. Put in the enclosure next to the Port Sydney bear, they huffed at first — and then began to cling to the fence and have what appeared to be more reasonable discussions through the wire.

The cubs settled in, both convinced they were as big as polar bears and as fierce as any bear ever was. They were still so small that their paws could reach through the wire mesh and they would try to grab anyone getting too close. We had to make the corridor out of bounds. They climbed the fence — the enclosure is roofed — they clambered over and through their hollow log — they wrestled with each other in the evergreen we keep in the cage for them so that they can have privacy. When they had nothing else to do, they boxed each other, pushed and roared — sounding like a murder in progress, though there was never any sign of blood. When they were hungry, they roared even louder. I rather think the Port Sydney bear regarded them as very ill-mannered.

Three bear cubs are easily managed, though, with all the other creatures needing care, we were busy. However, we had actually taken time to sit down and talk to some human friends who had come to visit the Sanctuary. We sat in the wide doorway of the barn, out of the sunshine, enjoying a cool breeze. The telephone rang.

The call was from Winnipeg. The reason — another orphaned cub. Could we look after it until it was old enough to go free? Yes, they would send it down. Yes, they would find a release place for it next spring — in Manitoba. Yes, we would take it. Three cubs — four — not that much more work. Fine, they would make transportation plans and let us know.

Before she ever came to us, we heard a great deal about her on the radio and read about her in the newspapers. Apparently a press release had been sent out from Winnipeg:

> An orphaned black bear cub found wandering east of Winnipeg has a new home. The eight month old cub is believed to have been left alone after a car hit and killed its mother a few weeks ago. The cub will stay at the Assiniboine Zoo in Winnipeg until final arrangements can be made.

Bit by bit the information filtered through to us. The cub was female — rather nice because the other three were all males. She weighed thirty-five pounds — was eating voraciously. Like most cubs, she huffs, puffs and swats in an attempt to be fierce.

Again, Bear Skin Airlines provided transportation — and again our van manoeuvred its way through all the miles of road construction on Highway 11, between Huntsville and North Bay, to arrive at the airport five minutes ahead of the plane. The white cage was made from oil drums and contained one very small, irate bear cub.

Bear Skin Airlines deserve all the credit that can be given. They had transported two cubs, willingly and free of charge. The calls kept coming — three more bear cubs from Winnipeg. And this was the second week in September — the weeks the terrorists erupted into our headlines, and stopped all air travel. Three cubs waited in Winnipeg until the grounded planes could fly again — and waited. When permission was given that planes could fly once more, space was at a premium; one by one the bears were flown to North Bay. Three times our van made the long trip.

Seven bears. In previous years, we had handled more than seven easily enough.

Then we had received a call from the Ministry of Natural Resources in Thunder Bay asking if we would undertake the care of a cub from Sioux Narrows. In the spring, she had been found, an orphan cub, and kept, more or less as a pet, at a camp. While the cub was small and as cuddly looking as a teddy bear, she had been fed scraps and encouraged to sit at the picnic table and share lunches — and, no doubt — pose for "sweet" pictures. Then, suddenly, she was too big and had become aggressive. She, being a real bear, was not quick to pick up polite manners. Would we take her? Since they had no idea just how the trip from Sioux Narrows down to Rosseau was to be accomplished, we called in the help of a Sanctuary volunteer who lives in Thunder Bay. When I asked him if he could pick up the cub, he did not try to explain to me that Sioux Narrows was a good six hour drive from his place. He said, simply: "Of course." He drove to Sioux Narrows, collected the cub, brought her back to Thunder Bay and put her on that most blessed Bear Skin Airline. Again, Tony met the plane in North Bay. This cub was dark, dark brown and blind in one eye.

Then the phone began to ring — and ring.

Toward the end of October, every year, the Sanctuary sends out a newsletter to all those very necessary people who support our work. The letter goes to the printer around the middle of the month. This year I cheerfully stated we had sixteen bear cubs to care for all winter, as well as the rest of the animals here, and also inferred that we would be working hard. Though the cubs had begun to arrive quite regularly, we still had not fully realized the extent of the problem facing us — and the bears.

After the letter was printed, I made an enclosure to update the situation: the insertion stated we had twenty-four cubs. Before the newsletter, with the insertion, was finally mailed, I had to update the insertion — we had thirty-five bear cubs. We received tremendous support. Still, by the time we were sending out the receipts for the contributions that people had made, we had to update again: fifty bear cubs.

We finally clued in: Ontario was facing a problem — and officially doing nothing. The volunteer who had assisted in the rescue of the Sioux Narrows cub sent us an article from the *Thunder Bay Chronicle–Journal* which helped to explain the situation in that area.

NUISANCE BEAR PLAN COMING ALONG

Nuisance bears will continue to be a nuisance in Thunder Bay for the time being. Thunder Bay police and the Ministry of Natural Resources are still ironing-out details of a proposed nuisance bear protocol which would state just who handles calls and how. The protocol is "in progress" . . . The Ministry of Natural Resources Thunder Bay District enforcement supervisor said yesterday, "I'd hate to put a date on it right now. We're slowly plugging away."

Very slowly. The cubs were orphaned and dying. No waiting for the Government to "plug away". The member of the provincial parliament for that district confirmed the official attitude.

"If you call the Ministry of Natural resources, what you are likely to be told if you are a private land owner is to shoot the bear, or call the police and have them shoot it. There is no strategy here to deal with orphaned cubs."

Unfortunately the situation was similar throughout all of Ontario. Nor were we successful in capturing every cub which was reported to us, and we knew that, if the cub was not found, it would not survive the winter. A cub would be sighted at a bird feeder, on a porch, at a dump — or simply wandering in a nearby field — or, most frequently, up a tree. We would respond, set the live-trap, wait — and the cub would have wandered on so that it was not found. Sad, and discouraging. So we do know that while so many little cubs are sleeping in our enclosure, even more are starving out in the bush. They will finally go to sleep. And, in all likelihood, not waken in the spring.

As the cubs continued to arrive at the Sanctuary, our financial resources were stretched to the limit. The media, however, not only sensing a good story, but also, in some cases really caring about the cubs, gave the situation consistent coverage. One radio reporter also interviewed the local Ministry of Natural Resources biologist. The biologist expressed concern that the cubs might become accustomed to human handling and so might cause problems when they were released and suggested it might be dangerous to try to help them. Did the Ministry know what to do, then? The biologist paused a moment before she answered: "No, we have no plans." That interview was at the end of December.

The official answer to every orphan situation is, "Oh — let nature take its course." As though guns and cars and garbage left out were all a part of nature! The drought — yes. When food is scarce for deer, the Ministry of Natural Resources establishes feeding stations. Evidently, those feeding stations are a part of nature's course. But for bears?

The cubs arrived from all across the province. The local Parry Sound office of the Ministry of Natural Resources grumbled and did nothing. MNR offices from Sudbury and North Bay, from Pembroke, Kingston, from Timmins and Moosonee all sent cubs. We met Conservation Officers and transferred cubs from their cages to ours in the back of the van — in Algonquin Park, North Bay, the French River. Officers drove up our lane with cages in the backs of their trucks. Only the city of Ottawa seemed to realize that each cub would cost us about one thousand dollars to raise and release — the government had not one cent to spare for the orphans.

Never mind! The people cared!

Each cub had its own story. Each cub was special — however, some stories stand out.

The Moosonee Bear

Moosonee, on James Bay, is a long, long way from Rosseau. Away up there, a very small cub was orphaned when her mother was shot for being a nuisance. The cub began to haunt the environs of an Indian Healing Lodge near the town — the natives called the police. An officer (whose wife was from Rosseau) managed to live-trap the cub — and phoned us. Then he made arrangements for the bear to come down as far as Cochrane on the Polar Bear Express. Since the hunting season was in full swing, the cars on the train were full of hound dogs. However (this is a restorative to one's faith in human goodness), they put an extra car, empty, on the end of the train and the little bear had a quiet, private ride all the way to Cochrane — where Tony met the train.

A Very Hungry Bear Who Looked for Food at a Dump

As you no doubt know, for many, many years wild bears have found food at the dumps where humans throw out food they don't want to eat — a very careless thing to do. Still, the bears knew where it was easy to find food. And in the summer, lots of humans go to the dumps to watch the bears. However, if some think the bears are a bother, they may shoot them. Out near the little village of Otterville is a very beautifully kept dump and a keeper who does a very good job, and also cares about bears. So when a very, very small cub started coming in for food — he watched carefully and when he knew that the cub had no mother — he phoned the Sanctuary. So we took out a live-trap and he baited it. He said: "I'll catch her even if it takes weeks because she is so small she just won't make it through the winter!" He knew what her pathway into the dump was and he knew the pine tree she climbed up at night to sleep. He took the trap to the foot of the tree, put some very tasty treats in it, and left it there. The very next day, the little bear went into the trap.

The dump keeper phoned happily: "We've caught her!" And so we brought the little bear to the Sanctuary, where she doesn't have to go scrounging in a dump to eat. And the keeper of the dump phones quite often to see how his little bear is doing.

The Little Bear Who Supervised Golfing

Around Muskoka there are many large, lovely hotels, surrounded by green lawns and gardens and, almost always, a golf course. The people who go to these hotels are usually from the cities and don't understand that if you leave a bear cub alone it will not hurt you. But they had heard that an orphaned cub cannot survive the winter all alone. So, at a big hotel on Lake Joseph, a small cub was seen up a tree on a golf course — and no-one thought it was really interested in golf. But it wouldn't go away — and so they phoned us to ask if we would come and take it away.

Since we know that cubs really are not one bit concerned about playing golf, we did take a live-trap and set it down near the place where the cub was often seen. One day , then two, passed — and then the cub, very hungry, smelled the apples which we had used for bait in the trap. In he went, the door closed behind him and he was safe for the winter. We brought him to the Sanctuary, gave him a warm house deep in clean straw, lots of water to drink and kibble and apples to eat. And he had a couple of other cubs for friends. He ate and ate, and from being a skinny little cub, he became quite fat — which is what cubs must be in the winter so that they can hibernate and wake up fine in the spring — to go free.

The Little Bear
Who Shared Breakfast with the Squirrels

We first heard about this little bear when he appeared at a new golf course which was being built. The men there phoned us but he went away before we could catch him. A huge, huge highway was being constructed nearby — he must have crossed that. A day later we had a phone call from some people on Rankin Lake Road who said that a little bear was up on their porch eating the food they put out for the squirrels. We took a live-trap over. The house was in the woods and the porch had to be reached by a high flight of stairs with a twist in the middle. Up there, sure enough, was lots of squirrel food — and some piles of bear scat which showed that the cub had been there. So we set the live-trap and baited it with apples — and, in one day, we had caught the little bear!

The Little Bears Who Ate in the Crab Apple Tree

The people who lived in this neighbourhood were very concerned about the little bear. The houses were on a hillside, with woods out behind, and each house had a lovely green lawn and gardens and trees. One house had a tall, tall oak tree beside it, with a little crab apple tree on the lawn. The little bear was sleeping every night up in the oak tree and feeding every day on the apples from the crab apple tree. However, he was so very small and so very thin that the people were worried about him and so they called us. We took a live-trap and set it, baited with sweet things, under the apple tree and we caught the little bear.

She was quite sick with pneumonia and we were afraid she would not live. Tony put an intravenous feeding tube into her front leg. For three days she was unconscious. Then she began to improve and started to lap up Esbilac (which is a type of drink one gives to wild animals because ordinary milk is not good for them) and, after a week, she was much, much better. In the spring, she would be able to go free.

Sanctuaries are not elastic — they do not automatically expand to accommodate any number of creatures. We had a good deal of construction and adjustment to undertake so that we could take care of all our bears. On the east side of the valley, beside the rocky hill, Tony built a new, large enclosure which would accommodate six bears. Each was given a hollow log, all covered deeply in evergreen branches. The largest of the cubs were put out there — the Port Sydney cub and the Manitoba cubs. When the cold weather came, they hibernated.

The pen on the west side of the barn was divided into two; the pen downstairs in the barn was also divided into two. Again, each cub was given a log or barrel and lots of evergreen. These cubs were smaller and still needed to be fed; some slept around the first of January, others remained hungry. Tony built another enclosure upstairs in the barn. This one housed a complete evergreen tree! The bears liked it. Three of them slept but the fourth, being smaller, still wanted food. (He actually managed to escape through an impossibly small hole under the door — but he didn't get far. Lynda and Janet saw him, caught him and put him into an empty enclosure by the barn — so the other three finally were able to sleep.) Five of the enclosures on the east side of the barn had three cubs each. Then, so that we would have a spare pen, a new one was built for Puzzle, the fox. He was moved out — the pen was empty until the "escapee" was secured therein.

None of this work would have been possible if it had to be done by Sanctuary staff only — we are always amazed and very pleased by the number of volunteer workers who come by so willingly and show their care for the wildlife by physically helping in the day-by-day slugging, which isn't glamorous, but is so necessary.

And food. The kibble for all our animals has always been donated by a pet food company. Never before had we used so much of it. But apples! Apples by the bushel — seven or eight bushels a day. Some were donated, most we had to pay for. Bushels of apples lined the hallway downstairs in the barn. Of course, the beavers, the porcupines and the deer like apples, too. Last year we planted an apple orchard but it will be many, many years before it will ever supply the number of apples we used that fall.

We had thought that maybe the flow of cubs had come to an end. For almost a week the telephone had not rung (about bears, anyway) — but then, one night, just before midnight, the Ontario Provincial Police called. Away out Brunell Road a little bear cub, hit by a car, was lying beside the road, very badly hurt — could we come and help it? Tony and Janet took a small cage and lots of blankets, and set out through the darkness. The trip was an hour from the Sanctuary but the policeman waited beside the unconscious bear. It seemed to be dead but Tony felt carefully and detected a faint heart beat. They wrapped the bear in the warm blankets and brought him back to the Sanctuary, into the warmth of the nursery, then left him to the quiet.

For three days the cub was unconscious and we worried . . . but, then sign of returning life! — if he was touched, he would growl softly. Most likely he had a terrible headache. When he could finally lift his head and look at us, he was still very afraid, and we did not want to stress him further by touching him. For several days he would neither eat nor drink — and then, just a sip of water — and, finally, a real drink. One day, just to tempt him (temptation isn't always bad!), Tony put a little square of chocolate in the cage. No bear could ever resist sweet things. He ate it. We tried chocolate chip cookies. He ate those. And chocolate pudding — into which we were able to slip his pills. And, when his appetite was finally aroused, being quite aware that no wild animal should grow accustomed to sweets, he was offered apples and kibble. These he now ate willingly, and all that was left of his few days of dissipation, was his name —

"Chocolate", "Cookies" or "Pudding", whichever suited best at the moment.

Finally, on Christmas Eve, the snow came and the weather began to grow cold. One by one, the cubs went to sleep. Only three or four very small ones still had to be fed.

And we thought the season was over.

On the morning of January 3, 2002, the phone rang once more. A small cub, about thirty pounds, was seen up a tree in Eganville. When the Ministry of Natural Resources was contacted, the officer said: "Leave it alone — it'll be alright." What nonsense. It would die. The Ottawa Wildlife Centre would try to rescue it. And send it here, of course.

January 5, 2002, I found a message on the machine. "There's a bear cub on my porch and I don't know what to do. It's very small."

I phoned Tony and relayed the message. He phoned, but the cub had disappeared. However, he told the lady to leave out apples — and, if the cub returned, to tell us. About ten o'clock in the morning she phoned — the cub was back on her porch. Tony had to repair the trap — the last cub captured had done his best to trash it — and then, Ed and Janet drove down to Port Sandfield, where the cottage was by the lake, and set the trap. Janet said that the cub was up in a pine tree.

All day the snow came down, thick and fast and damp. Around three, the phone rang again.

"I've caught him! He's in the trap!"

He was put into the pen out by the barn — along with the little fellow who had tried escaping. We thought he was small — this bear weighed twenty-five pounds at the very most. He ran around the cage — with the "escapee" monitoring every move — but grabbed up most of the kibble when he passed it. They eventually settled and maybe even kept each other warm.

He was number fifty-four!

18 – Sydney Bear's Springtime Release

Springtime at Aspen Valley inaugurates a new season of excitement and great activity. Creatures who have spent the winter asleep in deep hibernation begin to stir and become restless to explore the exciting world of trees, meadows and streams alive with new adventures.

Release of 54 bears (the situation in the spring of 2002) presented an almost impossible problem. A few of them would not be ready, because of previous injury problems, for immediate release; they would still require further care at the Sanctuary until they were able to cope on their own in the bush. The Manitoba bears would be flown back to their own province. For all of the others, the great majority, careful planning was required.

As with the release of all wildlife creatures, maps were carefully explored to determine release sites as far from human habitation as possible and in locations that would provide adequate food sources. Transportation plans were worked out It would be necessary to arrange for the bears to be tranquilized in preparation for travel.

An exciting new part of release planning in 2002 was the equipping of each release bear with a radio transmitter which would enable the tracking of each animal for a two year period after its release. Such tracking equipment, provided for research purposes through Cambrian College in Sudbury, Ontario, would enable us to see how each bear survived in the wild and how far it might travel.

In the chapter entitled "An Overbearing Situation" you read about an orphaned bear cub found near Port Sydney:

> Sometime in early May, near Port Sydney, a mother bear was killed on the highway. Shortly afterward, a very tiny cub was seen in a Port Sydney neighbourhood, starving, lost and, without some help, doomed. Tony and Janet took a live-trap over to River Valley Road.

. . . So the first bear of the year came to a small enclosure at Aspen Valley. He would retreat into his log and charge out again — huffing mightily — Except that the little black bear cub weighed only ten pounds. The Port Sydney cub had been orphaned by a road accident. We knew that he must not be tamed, that he, if he was going to survive when he was finally given his freedom, would have to be wary of humans. Still we watched him, enjoyed him, and Tony fed him.

The Sanctuary is very, very fortunate to have the support of volunteers who come and work hard. One of these volunteers has come for several years, is as willing to clean cages and wash dishes as they are to do the more fascinating work of actually contacting the wild creatures. Willingness to wash dishes is a major asset. This particular volunteer knows a great deal about animals. She has her own team of husky dogs — most likely the best cared-for huskies in Canada. She also teaches a class of grade six students at V. K. Greer Public School in Port Sydney. And she told her students about the cubs, especially about the cub from Port Sydney. The children were disturbed about the plight of all the cubs, including the one from their own community.

They decided that his name should be Sydney and that they should help to support him. They came up with ideas.

If you had visited the school that fall, the first item you would have seen as you opened the first door would be a colourful sign made by the class:

>Visitors are Welcome to Make a Donation to
>
>Operation Adopt-a-Cub.
>
>The donation jar is in the office

Beside the jar in the office was a pamphlet announcing some of the coming activities:

> We will be having a cupcake sale on Thursday the 29th. We will then be having a teddy bear and tee-shirt draw. There will be a colouring contest, popcorn and hot chocolate sale.

One wall in the foyer was dedicated to bear information. Several posters explained the situation:

> Why Are There So Many Cubs?
> What To Do If A Bear Comes To Your House!
> How Did the Port Sydney Bear Come To Aspen Valley?

Information was given about bears and other forest creatures under stress that season. At one side were lots of bear cub pictures, especially of the Port Sydney cub. On the other, a thermometer — from "0 to 500 dollars" — with bear paw footprints climbing up as the money came in.

When I visited the school before Christmas, the paw-prints had already climbed to 400 dollars, and I was treated to a song the class had written about their bear cub. They sang it well — and with very sincere enthusiasm! (Ms. Turansky's class made this song for Sydney).

P. SYDNEY URSUS!

> Who's got a "V" that's long and white?
> Sydney's got a "V" that's long and white.
> Who hibernates on a cold winter night?
> Sydney hibernates on a cold winter night.
> Winter night, "V" that's white,
> Must be Sydney, must be Sydney,
> Must be Sydney, Sydney bear.

Who's got paws and a suit of black?
Sydney's got paws and a suit of black.
Who has big ears on his head?
Sydney has big ears on his head.
Ears on head, suit of black
Winter night, "V" that's white,
Must be Sydney, must be Sydney
Must be Sydney, Sydney Bear,

Who's got a big wet black nose?
Sydney's got a big wet black nose.
Who shows attitude, "GRUFF, GRUFF, GRUFF!"
Sydney shows attitude "GRUFF, GRUFF, GRUFF!"
"GRUFF, GRUFF, GRUFF", wet black nose,
Ears on head, suit of black
Winter night, "V" that's white,
Must be Sydney, must be Sydney,
Must be Sydney, must be Sydney bear.

Who very soon will snooze in the hay?
Sydney very soon will snooze in the hay.
Eight little cubs too tired to play.
Sydney and his friends too tired to play.
 Tired to play, snooze in hay,
"GRUFF, GRUFF, GRUFF, " wet black nose,
Ears on head, suit of black
Winter night, "V" that's white,
Must be Sydney, must be Sydney,
Must be Sydney, Sydney bear.

Eventually, after a long, cold winter, spring began to arrive. Sun shone brightly, warm breezes melted the snow and ice. And our 54 bears, including Sydney, began to stir and waken — shaggy and very hungry. Ms. Turnasky, of course, told her class what was happening and indicated that soon Sydney and the other Aspen Valley refugee cubs would be given freedom in the bush somewhere. The class decided that they would like to visit him before he left.

On behalf of Sydney, I wrote a letter to the class. I confess to having used anthropomorphism in that letter. "Anthropomorphism" is such a huge word — it refers to the practice of attributing to animals the characteristics of humans — the feelings , attitudes, words and actions of humans.

This is a common practice in films, television, books, songs and plays — especially so in Disney productions. When it is done it is probably because we know so little about the true attributes of animals which may be far grander than when we belittle them by portraying them as though they were human. However, having declared what I believe to be true, I still use the practice at times with children, simply because I can think of no other way of not sounding too much like a biology textbook. The children of the grade six class of the elementary school in Port Sydney had really cared for the bear cub they had adopted. They had worked hard to support him and when the time came for him to leave for the free life they wanted him to have, they wanted to come to the Sanctuary to tell him goodbye. This is the letter Sydney wrote to the class:

> To all the young people in Mrs. Turnasky's class!
>
> "Well — I just woke up from sleeping all winter and found out that the snow is still not all gone and it isn't very warm — but Tony has been putting out big piles of apples and kibble in our enclosure — so at least I can eat. Am I ever hungry! All winter I have been sleeping with seven bear cubs from Manitoba. Just like me, they

didn't have any mothers, so someone sent them here. But now it was time for them to go home. A whole group of people came and stood around the enclosure and kept talking. But Tony had some apples and nice green grass so we went near to him, and a man with a long "rod thing" (a blow gun, I heard them call it), shot a little dart into our backsides. It felt like a bee sting, and we went to sleep. All the other cubs in my enclosure are being flown back to their home in Manitoba today, but I have been given a new enclosure all by myself, with a nest of evergreen branches to hide in and lots of straw and lots of FOOD!

Because I am hungry!

Tony says you are coming to see me, so I want to put on some weight before you get here because you will think I am skinny. I am. But I went for months without eating — so that's what happens. And my fur isn't nice and shiny like it was when I went to sleep — but you know what happens to your PJ's after just one night — so don't blame me! I can't even comb my own hair!

Anyway, by the time you come, maybe I will have eaten so much I will start to get fat and sleek again. I am anxious to meet you, because I know how much you have helped me and I really appreciate it all. And I want to hear your song.

Please come!

Love,

Sydney

Though they approached the enclosure very quietly, the bear was frightened. He had not ever been taught that human beings were to be trusted, but the children watched him, and said their goodbyes quietly. As they gathered at the front of the enclosure, they sang their song which they had written for him — very, very softly. And as they sang, Sydney slowly approached the front of the cage, listening. When they walked away, he clung to the wire, watching them.

Sydney went free! Then he wrote another letter to the children.

Dear friends:

I have spent my first night free in the wild — and it is wonderful! Dark and quiet — not a single sound of anything except the wind in the trees, the running water and sometimes an owl — no people, no cars — only the woods and the wild things. And I wanted to say thank you — because I know how much you cared and all the things you did, so that I could live a free life.

I was glad when you came to see me last week — a little bit scared at first because I had never seen so many people all at once. That was why I climbed up in the back of my cage and looked at you over my shoulder. At least from there I could look down and see all of you at once . . . we bears are usually quite afraid of people, you know. Sometimes people can be very cruel — but you weren't. You were very good to me. So I listened to you talking and began to realize exactly who you were . . . but it was when you sang your song that I really understood. You sang it so softy and gently. And so I climbed down to the ground and crossed the cage to be closer to you and to try to tell you that I was grateful to you, and that when I was living away out in the wilderness, I would not forget you.

You took a lot of pictures and Tony told you that I would be going free. Then you went away. I stood up against the side of the cage and watched you go. I watched until I couldn't see you any more.

This week a group of biologists came and they did things to us that I don't quite understand — we have something called a "radio collar" on and they weighed us. (I weighed 90 pounds — pretty good, eh?). Then they took us on a long, long truck ride. But at the end, they took us deep, deep into the woods and let us go free. Because I have this collar "thing" on, and a chip in my ear, I am going to be able to send messages back to you for two years. Then they will take the collar off — by that time, I'll be a big, big bear.

Please don't forget me and sometimes sing my song. I think that when you do, I will hear it, maybe faint and far away, over the hills and across the tree tops.

Thank you for my life,

Love,

Sydney

Sometime after the release of Sydney and his fellow cubs I received a bundle of stories that the members of the V.K.Greer Grade 6 class had written. They had tried to imagine what was happening to Sydney in the wild; their imaginations were following "their" bear and wishing him well. On the whole, they were very creative and delightful to read. Some of them were handwritten, others were neatly done on a computer. One of them was beautifully decorated with hand-drawn pictures — illustrations of Sydney, the trees and flowers in his new surroundings, and some of the wildlife creatures whom he met in his wanderings.

The stories illustrated how much of their thinking had been derived from television stories and movies, particularly those of Disney productions. In several stories Sydney went out into the woods, met a little girl bear, found a cave to live in, and lived happily ever after. Or Sydney made friends with squirrels or otters or beavers; his world became a wonderful school yard full of playtime and friends.

The best way to share something of those stories with you is to quote from them. Since I received seventeen of them (too many to include in full). I shall use excerpts from most of them to give you a gist of their content. I shall separate quotes from different stories by inserting dashes - - - between them; I shall also leave spellings as they were, for they were sometimes very creative (especially those shown in *italics*) !

Hi I'm Sydney. I lost my mom when I was little - - - It all started as I was walking across the *read* and the big metal [thing] came down and hit my mom . . . that was the last time I saw her but the last thing she said was "I love you so much". - - - Some people came and trapped me and put me in a *cage* and I get food every day. - - - I wanted to leave but they were helping me. I had gotten to a place called Aspen Valley. Later on I saw a whole whack of animals like raccoons, bears and a lion.

I was told that some people at a school called V.K. Greer were helping me. They did a whole bunch of stuff for me. I guess I was more famous than Winnie the Pooh. I was very, very happy for what they did. - - -

At Aspen Valley they were getting ready to release Sydney. - - - I'm member of the Aspen Valley Wild Life Sanctuary and now today I am getting released. They will still be looking out for me because they're putting a radio *collor* on my neck.- - -

It was a warm spring day. They had just given me my food, when someone drove up the driveway. It was Laurel Turansky, the one who suggested to her class that they should adopt me. She came up and started talking to *Audrie* in English. I can't understand some of it. Anyway, I think they were talking about where they were going to release me because I heard the words "Sydney" and "release *sight*" a couple of times. - - -

"Hold your horses Sydney, I'm a coming. With your dinner!" Tony hollered as he headed towards Sydney's enclosure, with a grin spread across his face. "Soon you'll be a wild bear and you'll have to find food and water for yourself."

Sydney looked Tony in the eye for a second as if to say thank you. Then he dug into his dog food and apples. Tony sighed and then went on to feed Siberia the lion. - - -

"Ouch! That hurt, don't poke that into me!" Tony was putting me to sleep. I didn't know but I would soon be released. - - -

"Come on Sydney into the truck, we can't take you to *you're* release place unless you get in the truck". "I'm being released *yay*!!

"All right Sydney out of the truck." - - - I was free to do what ever I wanted in the lush green trees and the deep promising forest of the evergreens. I was home! I ran swiftly across the moss on the wide meadow and the *dandy lions* tickled my nose and dyed them yellow with their pollen. Then I slowly trotted in the trees to find a place in the cool shade to take my afternoon nap. - - -

"I can't believe I'm actually out in the wild, but what do I do, actually I'm hungry I need my food dish, oh I forgot I'm in the wild *know*" - - - He found a huge bush of *rasberrys*.. He stuffed *have* of the bush in his mouth. Then he went to the pond and juggled the water down his throat. He was glad that he had everything that he needed but he thought in his head that he missed — Audrey, Tony, Janet, and Linda. So he had to find at least one friend to hang out with. - - -

After I woke from a peaceful nap a small racoon was washing a *crab apple* in the little stream in front of me. " Hey! What you doin?" I said , "Just washing an apple, yup, yup, yup" The racoon talked in a fast *squeky* voice 'Well what is your name Mr. racoon?" "Uh, I've never thought of a name for myself". "Why don't you have a name?" " I didn't know my mother so I don't have a name". "Hey I don't have a mother *ether* but my name is Sydney." Every thing was still after that. "Paws" he said. "Yup yup yup that's my name". " Okay then Paws, do you want to come with me?" "Yup sure I do!" "Okay then let's go!"

We had only walked about 10 minutes before we *smelt* a very foul smell that was coming from the shrubery in front of us. We wanted to know what was causing the smell because our curiosity got the best of us. "Lets go see what it is". I said. In an excited tone "yup, yup yup, I'm with you!" We walked over to the *sorce* of the stink. Then we saw a great big fat skunk with his out up in the air saying "Spray! Spray!" but all that came out was a bit of smell. "Hi!" I said in a friendly tone. "Hey what do think your *doin* near my stump! Now you have to pay the *prise*!" The skunk had a mean *sazzy* voice and didn't like us at all. Then Paws swallowed and said "What *prise*." "Umm let me see now. Ah ha I've got it you have to get me food for five moons or else I'll spray!"

The next day the old skunk made them get all the food he could fit in his fat little belly! Paws had collected all the berries that he could carry so he went to the skunks feeding trunk and *herd* him talking to himself. "Ah ha ha those two have no idea that I can't even spray. My plan is going perfectly!" The sly little racoon sneaked back into the bush to tell Sydney about the skunk. "WHAT!" "he's what! Grrr that makes me angry because I've starved myself for that little creep!" I was very angry so I stomped over to the stinky old skunk and roared so loud that that skunk ran all the way to the next forest! "Wow I never knew you had it in *ya* Sydney! Now let's go home."

"You know what Sydney *your* great!" - - -

About 3 or 4 days, later, I met up with my old friend, Oopsy, [who had been at the Sanctuary]. Oopsie had been released but I didn't see how. . . . I reached a place where another bear had been rubbing a tree. Although it wasn't Oopsie, I *smelt* something like him in the air. I followed my nose to a rock *over hang* with some saplings under it. "Oopsie!" I growled, "are you in there?"

A nice lady bear came out, followed by two cubs that looked identical to each other and they were pretty small. "May I ask who you are?" asked the girl "I'm Rosey". "Hi I'm Sydney. Oopsie might have told you about me and the sanctuary, well that is if you ever met him," I replied.

"Ya, he did tell me about you. What's that thing around your neck?" "What thing? I can't feel anything so I guess I don't know"

"Hmmmm!" I heard from behind me, I turned and there saw

Oopsie. He was fairly big; well he was bigger than I was, He dropped two huge fish from his mouth. "What is that thing around your neck?" "I don't know," I replied for the second time. "I don't think it is dangerous so we should just leave it."

"Okay." Said Oopsie with little concern. "Would you like some fish?" "Please, I'm starving!" I replied anxiously, trying not to be rude. Looks like *were* going to teach you how to catch fish." Said Rosey. "*Ya*", I replied.

After we ate the fish — which I enjoyed quite a bit — Oopsie and I went fishing. I was a fast learner but it looked like this was going to take a bit more time.

When we got back there was a gorgeous bear beside Rosey and she looked my age.

"Sydney, meet Buttercup, Buttercup meet Sydney." Said Rosey.

". . ." I opened my mouth but nothing came out.

"Hi." Said Buttercup. ". . . Hi" I finally said

"I'm Rosey's sister, we were orphaned when I was a little cub. My sister took care of me and then she met Oopsie and here are their cubs, Daisy and Rocky."

"I am an orphan as well, I was brought to the same sanctuary as the one Oopsie was born at."

You can *kinda* guess what happened in the next few years. Sydney's Release.

Tony attends his moose.

Larger Animals VI

19 – DAYS OF THE FIRST DEER

Not until after he died — long after he died — did the deer acquire a name. When he was living here he was the only deer, so we called him Deer. As the years have gone by we have raised literally dozens of deer and to refer to one simply as "Deer" was becoming confusing. To refer to him as "The First Deer" was awkward. Someone, and I plead, "Not Guilty!", named him, Bucky. He could not recognize the name, but we, at least, could identify the deer about which we were telling stories.

A fawn is one of the most beautiful of animals. Most creatures are born round and soft with eyes closed and huge mouths, as though trying to impress the world with the need for food. Some have no hair, some are softly downed, needing litter-mates to help keep warm — handfuls of life which need warmth, nourishment, care before they grow up to achieve the shape and wisdom of maturity. But a fawn is born beautiful. Small and slender, it's eyes large and dark and wondering, the ears large and soft as velvet, the body delicately brown with spots scattered like snowflakes, and a white tail which, carried high, flaunts its awareness of the world around.

In fact, the fawn is born a small, beautiful miniature of the animal it will become.

There! I have used the word "beautiful" three times in one paragraph — enough to emphasize the concept.

However, fawns are easy victims of the environment which we

humans have created: fast cars on highways, pet dogs running free, hunters. Humans who want them for pets. For the first days of its life, a fawn has no scent. The doe is able to say, "Lie down and stay!", and leave it while she grazes a fair distance away. Since it is scentless, and since the doe usually hides it well, a predator is unlikely to find it — at least a natural predator like a dog or a wolf or a wolverine. Often, humans find the hidden young and consider that it has been abandoned. They take it home and try to care for it; then they discover that raising a fawn is difficult, as well as illegal — and, often, phone the Sanctuary.

"We found this fawn out in the field. No mother around."

"How long ago?" If only a few hours have passed the best thing to do would be to return it to the place where it was picked up. The mother would find it — and hastily lick away the repulsive human scent!

Usually, however, the answer is: "Well, a couple of days ago." Too late.

When actual injury is the reason the deer comes to us, the most common injury is damage to the long, slender legs. Very fragile, the slim legs can be broken by cars or dogs, by a fall from the rocks, from being caught in a deep snowdrift. Deer are prone to extreme stress. Since stress can kill them in a short time, caring for a deer is not an easy task. Transporting a needy deer over great distances is dicey — but definitely worth the effort. And for those who do survive all the trauma — rewarding — for the deer, and for us.

Over the years, I have learned something about raising deer — certainly much more than I did when Deer (Bucky) came to the Sanctuary. Though badly injured, he was beautiful, special. While I was learning, I made mistakes. Rehabilitation was almost unknown then; no one really knew much about raising wildlife. I phoned conservation officers, and zoos and biologists. Most often I was advised to "put it down". Such a polite way of saying: "Kill it"!

Not until years later when I watched another three-legged deer we had raised bounding through the deep snow did I realize that, given his freedom, Deer would have been just fine.

I realize that a distinct difference exits between coincidence and miracle; sometimes I am not sure to which category some events belong. For instance: I had dropped in for a very infrequent visit to a friend with whom I had lived eight years previously. I had moved over to the property which has since become the Sanctuary. She still lived in the small house back by a river. We were both busy — but, once in a long while, managed to share a cup-of-coffee time.

I had been taking in raccoons and foxes and other smaller creatures for some years; my phone number was certainly in the Ministry of Natural Resources records. They had given me so many creatures that they most certainly knew the number for the phone I had answered most faithfully for the intervening eight years. So, I was somewhat surprised when the phone rang at my friend's place and interrupted our coffee time. My friend answered it, looked a little puzzled and said: "It's for you."

No-one but my dogs knew where I had gone visiting and they were not apt to give out much information. So, when the voice at the other end of the line said: "This is the Ministry of Natural Resources. We have a fawn with a broken leg. Do you think you could do anything for it?"

I said: "Yes", then couldn't resist asking: "How on earth did you know I was here?"

"Isn't this your phone number?"

"No".

"Well, I looked up the records." Old records. Old, old records. The fawn needed immediate help, desperately. Miracle or coincidence?

I returned home to await the arrival of the Ministry truck and the fawn. And I thought about three-legged animals: about Duffy, the raccoon, who had thumped around the barn for six weeks, leg in a cast, and managed to climb to impossible places from which I had had to retrieve him, but who had never admitted a handicap. I also knew race horses, when similarly injured, were often destroyed. I thought: "That decision will be up to a veterinarian to make" and then thought: "That's not fair — my decision!" But it was a decision I could not make until I had seen the fawn.

But I didn't make it then. The green Ministry truck drew up the lane and stopped. A cardboard box was in the back. I went to see it. The Conservation Officer, whom I did not know, lifted it down to the ground and opened it. Clearly, he was as uncertain as I was of what the proper procedure should be. All I saw was a young fawn, curled up. And bloody. Blood everywhere.

"Dogs", the officer said. "Damned dogs running free."

"Damned people". I couldn't resist correcting him. "who let their dogs run free."

A back leg was twisted unnaturally out behind, blood covered. The fawn lay very still, not reacting to either the sight of us or to our voices.

The officer shrugged his shoulders. "That leg's chewed pretty bad. We could just cut the strip of skin and it would come off".

"No", I said, not sure the fawn would survive that trauma. "I'll take him to Dr. Christie".

Though the evening was closing in, and I knew the veterinarian clinic would be closed long before I could drive to Parry Sound, I phoned. "I have a fawn. A couple of weeks old, maybe. It has been mauled by dogs." I did not want to say: "It might have to be put down". It had moved its head and was looking up at me, and his eyes were large and dark and shining. "Bring it in", the veterinarian said, a hint of weariness in his voice; his day should have been ending.

As I drove in, the fawn was in the box on the seat beside me, his head nodding weakly, his eyes closed now. I wondered how his day had begun; the dew-wet dawn, the sun rising over the tree tops, lying close to his mother (the warmth of her body against the morning chill). Her milk, plentifully flowing, filling him with life, the sun beginning to dapple through the green ferns, and the morning chorus of the birds. For awhile he had been with her, tentatively nibbling the wild raspberry leaves. Then she would have nudged him to the shelter of low, covering bushes and tall grass. As she moved away, grazing, he would have slept.

He would have been asleep and alone — and then — snuffling noses and sudden yelping. He didn't yet understand about dogs . . . he might want to run but would stay still because his mother told him to be still. He would know spasms of terror as the yelping dogs grew nearer and closed in. Maybe he did try to rise and bleat for his mother, but it was too late — the dogs closed in on him and began to tear at him. After that, maybe, he didn't know much until he found himself tied to a tree, and the dogs were gone, and the men were standing around looking at him, and all the blood.

He would have called for his mother, but, quite evidently, she could not come.

Away back then, Dr. Christie's hospital was on a road just outside Parry Sound, a road which curved away from the highway, out toward Georgian Bay. It stood on a large wood lot, surrounded by tall trees. By the time the fawn and I arrived, darkness was already closing in and the trees were moving in the wind. The lights of the hospital were on. Another car stood in the driveway.

Another emergency had arrived at the hospital — and the doctor was very busy. I had to wait. I sat on the rug on the floor, the fawn beside me so that its leg did not dangle. His great eyes were closed. The blood had clotted. Sometimes he panted for breath.

One hour ticked by, and two. The only sound was the wind in the trees. The only light, from the operating room, reflected against the darkness of the windows. The fawn was very still — and, at last, I heard sounds that seemed to indicate one operation was over. The doctor came out into the waiting room. He turned on the light. For a long while he stood looking down at the fawn.

I was afraid of what he would say. I wanted to make it easier for him. I said: "I think it will have to be put down".

After a long, long pause he said: "No. Let's give it a chance."

I do not know why so few, if any, veterinarians have been elevated to sainthood.

Dr. Christie, picking up the fawn, carried him in his arms to the operating room. Before I was called in the hands of the clock had moved passed eleven p.m. The fawn lay on the table, clean, breathing evenly. The leg had been amputated high into the hip and the skin folded over the wound, rather like an envelope, and stitched neatly into place.

I had expected he would simply clean off the place where the leg had been chewed off. "Why did you take it off so high", I asked? Somehow, asking a question was simpler than merely singing, "Hallelujah!" because the fawn was still alive.

Two reasons", the doctor explained patiently. "First, a dangling stump would knock against the other leg and cause sores. This way, the remaining leg is free to move toward the centre and make balance easier". Both reasons seemed logical enough, and, as time went by, proved to be correct.

For a few moments, we watched — the wind outside, the moving trees, the fawn's breathing were the only sounds. And then, the breathing stopped. "Dr. Christie", I whispered, "it just died."

He took the fawn into his strong and capable hands, lifted it and thumped it down hard on the table. "No, you don't!" he thundered at it. And the fawn did not die. He didn't dare! He breathed again. "And you", he stated, looking directly at me, "will sit up with him and watch him all night. If it happens again — you do exactly what I just did. Understand?"

I understood.

By the time Deer and I were home again, not much of the night was left. I put Deer on a warm blanket in a box beside the couch. I lay with my hand touching him. Nights in the country can be very quiet, yet the quietness of the hours when life and death are in contest has an intensity which is almost physical. I was conscious of the surrounding wilderness — that somewhere out there perhaps a doe was still calling for her fawn. Twice he stopped breathing. Twice I thumped him. Brutally hard! The darkness outside gave way to light and the fawn lived.

With the dawn, I gave him his first bottle of warm goat's milk and he drank eagerly.

Because of his recent experience with dogs, I had expected the presence of four dogs in the house to be a problem; would they terrify him? However, Kate, a Labrador/cross, put her head into his box and gently licked the leg wound and all the teeth marks which the dogs had left around his neck and ears — as gently as a mother doe would have done. He lay very quietly and accepted her administrations as though he understood that she cared for him. For the first days she watched over him carefully.

Then, he wanted out of his box. I took him outside to the grassy dog-run. There, he wobbled around on his three legs, learning to walk and then to run. And he was guzzling goat's milk as though he had never had any other kind.

Back when Deer entered my household, the Sanctuary was not really part of the plans for the future; in the world of wildlife, rehabilitation was merely a distant rumour. The valley was still designated as a farm — the sloping, forest-surrounded meadows were classified as a grazing place for a couple of horses, a goat or so and a little burro named, Jacob. I had always wanted to live on a farm, and, as far as I understood then, my life on a farm stretched on into a peaceful, predictable future. At least, as far as life with goats and burros can be predictable. Though I had opened the door to the occasional skunk or raccoon, I had as yet no conception of what that meant to the future.

Now, in the present day of accumulated wisdom and with so much advice from all sorts of humans who are regarded as authorities (there, one must be selective!), wild animals are kept out in large pens and enclosures; they have little or no contact with domestic animals. Certainly, once weaned, they no longer live in the house. Better rehabilitation procedure, I fully admit — but it also means I have lost just a little of the challenge and excitement the mixtures of those early days generated.

However, away back in the "days of the first deer", I did not have enclosures and runs and cages for all the creatures which were given to me. Now I was faced with the problem of adapting my household to yet another unique inhabitant.

"He will not" — I announced firmly to the four dogs, the resident beaver and the skunk, who had just wandered out of the bathroom — "He will not live in the house. Deer do not."

They looked at me wordlessly, and probably smiled their secret smiles. The raccoon curled up on the couch, snored on. If he heard my announcement, he may have agreed happily. "No, of course not", and passed on to think of something more probable.

The house is small and I do make rather valiant efforts to keep it clean. Once I found a big Belgian horse which was boarding briefly at the farm (in the barn), with her nose pressed against the window. Chasing her away, I refused to notice the offended look in her big brown eyes. Again, amidst the general confusion of moving a washer into the basement, Bramble, the goat, gained access to the living room. I brushed past her and around her and pushed her out of the way for a good ten minutes before I realized that the creature standing on the rug was not a dog but a goat who was surveying proceedings from the top of the stairs. Only once did I actually invite the little burro into the house. We were having his first birthday party and it was fair that he should be invited to blow out the candle on his cake. He ate both the candle and the cake.

My dog Kate said nothing and continued to say nothing when Deer began to follow me into the kitchen and wait somewhat impatiently while I warmed his bottle. The dogs took it for granted that, on cool rainy evenings, he would lie with them on the rug before the wood stove. I said nothing when I found Deer and Duncan, another of my dogs, asleep side-by-side on my bed.

I often wonder who, out in the wilderness, makes friends with whom. Does a beaver ever willingly share his house with a wandering skunk? Do raccoons ever regard rabbits as anything more than dinner? I realize that, in the wild world, rabbits are prey to everything that eats meat.

Here, they are one of the best of companion animals, keeping all sorts of orphaned mice and squirrels warm in their comfortable fur. Only once did I have a problem rabbit. His name was Alfred. He chased the dogs unmercifully, nipping at their heels, pursuing them over couches and under beds. Eventually, he had to be exiled from the house. Deer, perhaps because he knew that the dog-run provided him with complete safety, was friends with everyone.

Apart from the dogs, his best friend was a beaver. When Deer was finally too big for the house, they slept together in a dog house. When Deer began to realize that food other than milk was very tasty, he shared his apples and lettuce and grain with the beaver. Raccoons, free to climb in and out of the pen as their fancy took them, often stopped to visit, touching noses with Deer and then plumping themselves down on their fat backsides, enjoying whatever food was in his bowl.

Jacob would come and stand at the fence touching noses, or simply lie down beside the fawn, with only the fence between them, drowsing in the warm sunshine. But Deer's greatest friend, a friendship which endured for their entire lives, was with Duncan, the Labrador.

Kate had, as much as she was able, taken the place of his mother. However, Duncan was his friend. Duncan, when very young, viewed life with wild exuberance. No Labrador puppy is ever quiet, but Duncan was irrepressible — huge feet, lanky bones and a heart brimming with passionate enthusiasm for everybody and everything. Deer was beautiful, delicate, still a little precarious on three legs, his hoofs dainty as dancing slippers. But the two of them played. Around and around in the pen they romped. They would play until they were tired, share a long cool drink from the bucket and then sleep side-by-side against the fence.

Of course, since I had decided that a three-legged deer could not be set free, Deer would need a much larger enclosure. I am quite aware that high school teachers are not underpaid but at that time in my life I had many demands on my resources and I did feel that since the Ministry of Natural Resources had given me the deer, they could also afford to give me sufficient funds to build him an enclosure.

And so I phoned. And I explained.

"On no", came the reply. "There's nothing in our budget for that!"

I pointed out that they were cutting new ski trails through the area — and new snowmobile trails.

"That is a different account. We have nothing for building enclosures".

I said several things. Eventually they sent three hundred feet of fencing — no posts, no help — but fencing. I had to add to it, naturally. But they did give it. In all the more than thirty years which have followed, there has been no repeat gift. Some while passed before I found someone to actually construct the enclosure. In the meadow beside the lane, it was a hundred feet by a hundred feet — lots of grass and bushes for shade and a pond. When I compare it to the enclosures which our deer have now, though at the time it seemed quite splendid, I realize now that it was not much more than adequate. But Deer seemed content.

My beaver at that time was accustomed to much attention from the public. I am almost ready to swear he recognized a camera when he saw it and was able to pose — anthropomorphic, I know, but it seemed true. When he was discovered by the media, Deer was small and spotted. His first appearance on television was fairly respectable. It presented a picture of a winsome spotted fawn, thumping about on three legs, looking large-eyed and slightly tragic into the camera lens. I wondered why I had no calls questioning my reasons for keeping such a sweetly pathetic creature alive — surely he was suffering. Just look into the depths of those eyes!

During his second television appearance, toward Christmas, when, in spite of falling snow, he had become very sure-footed, the camera simply stimulated his curiosity. Afraid of a stranger? Not Deer! He just wanted to know. He thrust his nose at the camera, at the equipment. Trustfully, the cameraman allowed the fawn to examine the camera, the tape recorder and all the wires which connected them.

Deer bit them in half.

And pursued the desperately photographing cameraman around the pen.

When the photographer had been in the pen for a considerable length of time, I finally opened the door and rescued him. "Did you get good pictures", I asked?

"No!", he replied, with barely controlled exasperation, "but I have a lens covered with nose prints."

A newspaper reporter had a somewhat better experience. Scouting for human interest stories, she sat in the living room trying to adjust herself to the fact that a skunk was curled up beside her on the couch. A cat and a raccoon were tangled up together, deep in slumber, on an armchair. A beaver, flip-flopping down the hall to meet her, she took in stride. Or tried to. Then Deer, pushing through the screen door, made his entrance. (Did he really know how to make an entrance?). He paid no attention to the beaver, who had gained access to my knee, totally ignored the raccoon, cat and skunk, and advanced to the reporter. Graciously, he allowed her to rub his ears and stroke his nose. He inspected her writing pad carefully, seemed to approve, and settled down at her feet, posing like a tired hound in an old painting , faithful at his master's feet.

Though the inhabitants of my house were fine and acceptable to me, sometimes visitors found life there uncomfortable. One night, a cousin of mine, a policeman who was quite accustomed to all sorts of challenging situations, came for a visit and had to sleep on the pull-out couch in the living room. In the morning, I heard a sleepy voice muttering: "What the — There's a ground hog in bed with me!" The ground hog was only a baby. Another ground hog was under the couch — it hadn't gotten into his bed. Waking up could have been even more confusing.

Another time, a female guest, to whom cleanliness is next to Godliness, and who considered my household in great need of redemption, complained somewhat bitterly: "But I can't take a bath! There's a heron in the bathtub!" Again, a frustrated philosopher, intent on conversing about the deep realities of life, in utter exasperation, stopped in the middle of a profound idea. "I cannot", he announced emphatically, "carry on a serious conversation with you as long as you have a skunk on your knee!"

One particular evening, I had actually managed to cook a meal for some rather courageous friends who did not mind eating under the hungry eyes of assorted dogs and one small deer. When the dogs had begged as much as they could manage, and retired, we finished our meal. Deer did not give up so easily. He saw food — left-over salad — on the other side of the table.

He did not consider going around the table. He tried to leap over it. A four-legged deer might have accomplished the jump. He did not. I cleaned up broken dishes. And the broken chair.

And even I realized he was much too big to be granted house privileges any longer.

One of the biggest disadvantages to trying to be a high school teacher in Parry Sound and a farmer (of sorts) on the Huntsville side of Rosseau, was the thirty long miles which lay between. I did not mind driving the distance. The highway curved between high rocks, tall pines and scattered lakes. I thought that if I ever grew so accustomed to the beauty of the drive that I did not appreciate it, I would have to take a season away from it and live in the city. Still, when part-way through the day the weather changed from warm and wonderful sunshine to pelting rain or whirling snow, those thirty miles seemed like an infinite distance. Though I did not enjoy driving in bad weather, my major concern was the animals left at the farm. Did they have sufficient refuge? Were they being drowned? Were they freezing to death?

I was still underestimating their intelligence.

"Before winter comes I'll have to somehow get Deer a place to live in the barn. Another stable or something." I said to a friend, who knew a good deal about wild things.

A condescending smile. "Ma'am, every deer in the country is outside all winter. Yours has a shelter and lots to eat. He wouldn't appreciate a stall or a barn."

Right. Still, one late November day which had begun sunny and relatively mild, managed , just after noon, to develop into a major blizzard. I stood at the schoolroom window and watched the snow whipping in from Georgian Bay. I felt panic begin within me; Deer was outside. He had never been in snow before. Were the bushes really sufficient to give him shelter? As I turned back to the class and watched those budding artists plying paint and brushes, I saw Deer, shivering, huddled against the fence, his eyes wide with fear.

I confess that I did manage to slip out of school a little before "Official Leaving Time for Teachers". As I drove, snow clogged the windshield wipers. The snowploughs were not out; snow clogged the highway. I could not see those rocks and trees and lakes. My little car pushed bravely through; I tried not to think of my suffering deer. I tried to concentrate on driving. Finally, through Rosseau, out the gravel road, a right turn and I was safe. If anything happened to the car, I was within walking distance of the Deer.

Once home, I swung into action. Snow drifted deep against the barn doors. I dug through. Under any circumstances I find a bale of hay rather heavy; that day I dragged bale after bale out of the barn, down the lane, through the blinding snow and then hoisted it above my head to tumble it over the fence into the run — bale after bale. My arms ached and my throat was sore with gulping cold air.

With deep interest, Deer watched.

After I had piled the bales to form the walls of the shelter, I carried an old wooden door down the lane to make a support for the bales of hay on the roof. More bales, and finally the shelter was finished, creche-like enough for a Christmas card, quaint and storm-proof, warm in the blizzard. I stood back and admired it.

Deer lay in his bed of snow which was building up around him like a cup and covering him like a blanket.

"Deer", I panted. "There, it's ready. Go in out of the storm."

He regarded me with dark and wondering eyes. He continued calmly chewing his cud. He did not move.

Not once, during the long winter, did he ever use his house. He knew, as all wild deer know, that if he lay down in the snow, and was still, he would be warm.

A fawn is always beautiful: to be truly handsome, a buck deer needs a set of splendid antlers.

I am not certain why Deer never quite managed to achieve this attribute; perhaps in that initial mauling by the dogs, some deformation of his skull occurred, or perhaps, because he lived in an enclosure, he spent time in the night darkness tangling his growing antlers in the fence. Whatever the reason, Deer never did grow a proper set of well balanced antlers. They always looked like confused weather vanes. When he was a yearling, the first little lumps, fur covered and showing no sign of anything irregular, appeared on his skull. It was as exciting as the appearance of a child's first tooth! He was indeed a Buck! The velvet buds swelled. Though I knew his first set of antlers would not have the magnificence of the trophies one sees on the calendar pictures of noble Scottish stags, I did hope for a pair of graceful spikes which would, when eventually he shed them, look well mounted on a plaque and hanging on a wall and labeled: "Deer's First Antlers".

At maturity, one antler attained the length of two inches and bent north like an arthritic thumb.

The other antler attained the length of about an inch and a half and wandered thinly west. However, first antlers are first antlers and I determined that regardless of their obvious defects, they would grace my wall. So, on a February morning when I noticed he had shed the smaller one, I massaged his head until the larger one fell off in my hand. Two small slightly bloodied round spots on his head marked where his antlers had been. Within hours, in the cold winter, those spots healed. But where was the other antler? I did want a complete set!

The following season when his second set of antlers began to grow, they showed more promise. Like straight and slender saplings, they grew gracefully from his skull. My hopes grew. At about six inches, covered with the soft velvet, the antlers sprouted their first point. One point north. One point west. However, they did grow about a foot in length.

And so Deer lived his given years. He had many friends — Duncan, the dog, of course, and visiting raccoons and skunks and cats. Many children knew and visited him and he greeted them all. He knew occasional cars which passed on the road; some he ignored; for the neighbour's, he might lift his head. For mine, returning from school, he would come to the gate and wait, ready to greet me, to be talked to and to have his nose rubbed. And to be fed, of course.

He never knew the freedom of the meadow and the woods. He never stood by a lake in the early morning when the mist was slowly rising, nor grazed on the new, green growth of spring. I wonder if he realized all that he missed. When, these many years later, I see a deer, handicapped as he was, surviving very well, free in the wild, I wish I had known and understood. But I didn't.

If, at some future time in some future world, I am wandering down in a meadow between two long arms of forest and I see a deer grazing quietly — if that deer, curious, raises a handsomely antlered head to look at me — and if when he does, I recognize Deer, I will apologize to him for never having given him his freedom.

Being Deer, he will look at me, some condescension in his attitude, and say: "Never mind. I had a great deal to teach you. It took my entire life".

Duncan likely will come bounding through the tall grasses. And wag his tail in welcome.

20 – CHEWY THE MOOSE

Nim-keas-quay, the bear, was not being cooperative. Of course, she did not know that the men who were trying to hustle her into the small carrying cage had wonderful intentions; they were going to put her in the cage, put the cage in the back of the truck, drive her a long, long way to a safe place in the wild and let her go free. She did know that she did not like or trust humans — and she was large enough to be aggressively threatening.

Four months before, when she had come to the Sanctuary, she had been completely helpless. Found lying beside the 169 highway, wet in the late winter snow, she had been so near death that a man had been able to pick her up in his arms and lay her in the back of his open pickup truck. During the twenty mile drive to the Sanctuary she had made no effort to move, and here, when he lowered the back of his truck, she had not even raised her head. I had carried her into the house, put her in a small cage and started to feed her goat's milk slowly, using a large syringe. She did not struggle. But she did take the milk.

Her recovery had been rapid. When we knew that she would live, she had been named Nim-keas-quay, Ojibway for Little Big Girl. She was moved outside to a small pen. As she grew, she was moved to a still larger pen, and by late May, she was a good-sized yearling, strong, aggressive and quite able to take care of herself in the wild. The time had come for her to go.

Maurice, her Ojibway friend, and another helper finally abandoned their efforts to lure her into the small cage. They built an enclosed ramp from the enclosure gate, up to the back of their truck and then into the cage — all Nim-keas-quay saw was the open enclosure door. She was out, up the ramp and before she realized what was happening was into the cage. The door to it was secured.

One angry bear, one bear with another reason not ever, ever to trust human beings.

A quiet seemed to settle over the Sanctuary; the only animals which remained were smaller creatures, creatures that could be more easily handled — raccoons, a few skunks, a fox or so, and, the beaver. I breathed a sigh of relief; large animals are wonderful but at that time I was caring for the Sanctuary pretty much alone, and large things — such as bears with huge grudges — could be difficult — even dangerous. So I fed all the small things, went for a quiet walk with my dogs and then home to bed — to a long, quiet night when I would not hear the fretful calling of an imprisoned bear and fantasize terrible scenarios in which he escaped and terrorized the neighbourhood. (I realize now that was pure fantasy — humans may resort to terror, but bears do not!). Tonight I would think about Nim-keas-quay, free again, moving like a dark shadow through the tall trees of a far wilderness, miles away from humans — a happy bear, forgetting us, free.

Quiet, I slept. Until the phone rang.

"Hello — that you, Audrey? Sorry to phone so late. This is the Centennial Animal Hospital. We've just had a moose calf come in. Hit by a car — but I think he will be okay. If we send him up in the morning, will you look after him?"

I heard myself say: "Of course" — with no hesitation.

That's how this particular relationship began.

Somehow, though, the nighttime quiet had been disturbed. A moose calf. How big was a moose calf? Did it take goat's milk, as the deer did? And how much milk? And where would I put it? And how big — how big — how big? Wondering, I finally drifted off to sleep.

And so the quiet night passed — one moment thinking about the little bear enjoying his first night in the bush and wondering if the thunderstorm which crossed Muskoka sometime around midnight was creating havoc also in that far away woods where he was camping— and the moose calf lying in the veterinarian's hospital down in

Bracebridge. Finally, just after dawn had pushed away the lingering storm clouds, I heard the approaching motor of a van coming around the bend and down the hill. The moose was arriving.

He was lying, wrapped in blankets, on the floor in the back of the van. Just in case he should suddenly recover from whatever tranquilizer was keeping him down, two young women sat near him, one on each side. A third was driving. She had driven rather quickly.

Only the head of the calf was visible. He was still partly asleep, but tried, groggily, to raise his head.

"He comes from highway 117", I was told. "His mom was there with another calf — but he was hit. He was just lying there and she went off. Just left him."

"The police came," another of the young women continued. "They were going to shoot him. But another man came along — he is a singer". (I am afraid I forget this name) "and was on his way to work. But he put the moose in his car and took him to Centennial." Good for him!

During my quiet night I had spent some time deciding just where I would put the moose. Fortunately, the large deer pen was empty — one corner of that was fenced off as a fawn pen. A small shed would give shelter, and a little creek running through would give water. A good number of fawns had grown up there — surely it would be sufficient for a moose calf. Temporarily, at least.

Very temporarily.

Together we moved the not-quite-awake calf into the warm straw in the shed of the fawn pen. I prepared a bottle of warm goat's milk. He was not quite ready for feeding but it seemed the thing to do. When the moose was safely settled, the van drove away and I was alone with the moose. At which point the phone rang once more.

The call was from the Parry Sound Animal Hospital. A very young fawn had been hit by a car — they had done all they could for it — would I look after it? Yes. But, I thought as I went to inspect the progress of the moose, the fawn would have to go into the barn. There I prepared a small pen.

The moose calf was awake and up on his feet; when I arrived, he was looking around the enclosure — speculating. I was to learn that moose do a great deal of speculating. I was to learn immediately that moose calves are bigger than fawns, stronger and a good deal more emphatic about their demands. The moose was trotting around the fence, pushing here and there, deciding how best to escape. I thought that, perhaps, if I fed him, he would settle down. Perhaps! He was certainly ready to be fed. I had prepared a beer bottle full of goat's milk, with a calf-nipple — he consumed it all in less than a minute and wanted more. While I prepared that, he ranged the fence. A fence which would not hold him for long.

Up in the barn was a large box stall I had used for my horses. What would hold a calm horse would certainly hold a moose calf — a calf as big as a week old cow calf — wouldn't it? If it would — how was I to move one agitated moose calf from the outside pen, up the lane and into the barn?

Help arrived. Dr. Ian White, a veterinarian from Parry Sound, drove up the lane. He was bringing the fawn. He did not know then and I didn't either, that his arrival at that precise crisis was the beginning of long years of his association with the Sanctuary. His first challenge was to be to help me move a moose calf — since then he has done everything from setting the broken legs of porcupines to giving a huge male bear a vasectomy. He found moving the moose no problem. He covered its head with a blanket and half-led and half-carried it up to the box stall. There, knee deep in fresh straw, and probably calmed by the semi-darkness of the barn interior, it settled down and slept. Leaving us free to tend the fawn.

A fawn is absolutely beautiful, probably the most perfectly proportioned, delicate, large-eyed of any baby creatures. The little fawn which Dr. White brought that day was curled up in a deep, warm box, on soft blankets — but it was so very badly injured that it did not live. However, it did manage to emphasize the contrast between a fawn and a moose calf. The moose tempts one to think that perhaps God is something of a cartoonist. The body of the calf is so very short and the legs exceptionally long and gangly, with knobby knees and gigantic hoofs. A tiny tail and no neck to speak of — but the head is huge and the big ears flop out of control. Big nostrils, big lips, and eyes as soulful as a cow's — not beautiful. But absolutely irresistible. When the moose is grown, it becomes beautiful, awe inspiring, majestic.

Alone with the moose, I began its care. I had a large calf-nurser which held a litre of milk. I tried to coax the calf to suck from the nipple. Since the last creature to use the bottle had been the bear, Nim-keas-quay, I had washed it and washed it and washed it; the little moose turned his head away, refusing it. Blaming the bear scent, I reverted to the beer bottle. He refused. Hoping that hunger would make him less fussy, I left him for a couple of hours. Presented with the bottle once more, he still refused. So, knowing he had to be very hungry and unwilling to let him starve himself any longer, I found a large syringe — the sort one uses to baste a turkey. I squirted the milk into his mouth. This worked! He swallowed the milk. Willingly. When we finished, he was covered with milk. I was covered with milk and the stable was well sprinkled.

But he had enough to satisfy him.

We cannot name every one of the dozens of animals who come through the Sanctuary; however, if the creature is to be with us for any length of time, it must be named. So, the moose — Bullwinkle was immediately suggested and rejected. Dudley was suggested and rejected.

T.S. Elliot wrote that the naming of cats was not only a difficult undertaking, but a baffling mystery. He wrote of that experience in a most excellent poem, a poem which has earned a prestigious place in the history of English literature. Had he been faced with the challenge of naming moose, what would he have done? Produced an epic of awesome proportions?

The naming of each moose calf which comes to the Sanctuary is important; one cannot refer to a calf that had been raised and released five years ago as simply: "that calf". If one does, a swarm of calf personalities come to mind, somewhat vague and non-specific. If one can say,"Remember 'Night Danger'", we can, and I do — remember who he was, where he came from and the details of his release.

We did have a calf named, Night Danger. The name originated from the signs along the highway, usually near a dip in the road, usually near a wetland where a moose would feed and perhaps cross the road in the night darkness. Actually I have sometimes wondered why we do not erect corresponding signs, facing the wetlands, to warn moose about thundering trucks.

Some years ago, we were given a moose calf which had been orphaned by a forest fire caused by a practice military operation at Camp Petawawa. Since he had a military background, we named him Captain. And, since the human who was his principle caregiver had other "interests", too, the name became Captain Morgan.

The local radio station in the Muskoka area is called The Moose. So, inevitably, we named a moose after the call letters: 100.1. Perhaps the most appropriate names we ever used were given to two little calves who arrived about the same time, and who were forever arguing with each other about who was to be given the feeding bottle first — Chaos and Havoc.

However, the first moose calf ever to come to the Sanctuary, long before I had learned the havoc and chaos moose calves can cause, was given a dignified Ojibway name. He was Chinneas: Little Big Man. All quite suitable — he came small, he left big. However, over the months he lived with us, the beautiful name grew shorter and shorter. As the magnificent animal stepped off into the wilderness, we said: "Goodbye, Chewy!"

This then is the story of Chewy.

The stable where he stayed was dim and hay-filled; the barn was quiet. He ate and he slept. He was alone. We did not want him to become accustomed to humans — the day would come when he had to return to the wild, where he could be hunted and killed. His access to humans was to be limited — but — he had been rescued on a public highway, taken by a well-known musician into the largest veterinary hospital in Muskoka. There he had been cared for by expert people from the clinic and the Humane Society. Of course the newspaper found out — and there, on the front page was a picture of Chewy, having been rescued, being carried into the van to be brought to the Sanctuary. So, the first weekend we were overwhelmed with visitors. We could only hope that those children, who were enchanted with the homely, wobbly little moose in the dimness of the stable, would begin to know that moose were something very much more than targets for hunters.

Watching the moose calf watching the visitors, I wondered what, exactly was going on in his mind. He did not respond to the repetition of his name, to their chirping or calling; he stood well back in the shadows, watching. Adults tried to coax him near; when he did not come, they told their own moose stories — everyone, it seems, had had an encounter, however brief or distant, in the woods with a moose. Children, stretching to peer over the side of the stable, watched him in silence. Perhaps they were puzzled or awed — few had ever before encountered such a small moose. But what was he thinking? What thoughts moved through that strange, huge head?

Even when he was calm enough and had grown somewhat so that we were able to move him back out to the fawn enclosure, his acceptance of humans was limited to only those who were actually involved in his care. This is a fact we have learned about most of the wildlife with which we are involved; they will accept their caregivers. The remainder of the human race means nothing to them.

However, his first day in the large enclosure seemed to indicate that even that connection was based on food supply. I had moved him into the fawn pen which itself was in the large deer enclosure. Because the enclosure had not been in use that year, the grass was tall — originally the field had been sown for hay and tended to grow shoulder high. When, early in the morning, milk bottle and syringe firmly in hand, I went out to give Chewy his first feeding, I opened the gate and went into the fawn pen. No moose. I searched the entire pen. I looked into the little shed.

Absolutely no moose. The fence around the pen was only five feet high — so I began to search the large deer pen. The fence around that is a good ten feet high — I searched, I called — no moose. Finally, in desperation, I summoned the services of our Border Collie, Laddie.

Laddie was very familiar with wildlife. So I said to him, "Somewhere in this enclosure, somewhere in the long, long grass, is a moose calf. A hungry moose calf. Find him."

Laddie did. Within a couple of minutes he stopped ranging through the enclosure, looked up at me, his tail wagging slowly, and woofed — while Chewy struggled slowly to his feet. When he saw the bottle and syringe, he trotted over to me, and condescended to allow me to feed him. Finding him, with Laddie's help, was difficult for only a couple of days. Then he knew that, when I called, six times a day, I had food for him. The long grass parting like waves on each side of him, he would come trotting, making little mewing sounds like a starving kitten. Ridiculous, I thought, out of a mouth as big as that.

Soon, he wanted more than merely milk. That year in Muskoka was a year when the tent caterpillars invaded, stripping certain trees for miles around. Moose, like beavers, love aspen — but nowhere was aspen to be found. Fortunately moose, unlike caterpillars, do like willow. Once a day an assistant would take the truck, cruise along the back roads and cut willow branches — a truck full for a baby moose. When we dragged them into the pen, we would try to stand some of them up so that they might look like trees. The illusion made us feel as though we were doing our best to give Chewy a natural life; Chewy himself didn't notice. He munched happily.

Sometimes I wondered why one could not see the actual unfolding growth of the moose calf; it happened so rapidly. Though the almost comic proportions remained unchanging — his head was always too large, the neck nonexistent, the legs too long and the body too short, he grew. The hump on his back began to develop and the beard beneath his jaw. He also began to develop a mind of his own.

Once again, when answering the telephone, I had heard myself giving an enthusiastic "yes!" to a commitment of two years at least — a commitment for an adventure I had never before experienced: six tiny Eastern Timber Wolves had been orphaned. Would I raise them? Pat Snell, the woman who had been giving them care when they were two days old, would keep them until they were weaned and then — well, she lived in the city, and that was no place to raise wolves. We started to prepare; at that point finances did not permit an appropriately large enclosure. That, I prayed, would become a reality before the wolves were large enough to need it. Meanwhile, the enclosure was about twenty feet square, with very high board fencing on three sides and wire on the fourth, lots of straw and logs for playing and hiding. As the pups were growing, so was Chewy. Steadily.

Not long after the wolves were safely in their enclosure, and romping through each day with a singular joy which wolf pups seem to know, Chewy gave us a lesson on the order of things in an ordered world. Because they were the youngest, we fed the wolves first — and stayed to play with them for awhile. Wolves, like beavers, are family animals and need time and attention. Waiting, Chewy paced in his enclosure. Finally, a Sanctuary assistant, took the truck out along the back roads to reap his daily harvest of willow branches. Home again, he piled the branches into Chewy's enclosure. From a distance (sulking?), Chewy watched. The assistant began to prop some branches upright against the fence. Then he heard the thud of galloping hoofs — Chewy was charging him — with no playful intent. No one argues with a half-grown moose. The assistant exited the enclosure quickly. And we learned that one did not enter the moose pen smelling of wolf pups. After that, when we realized that moose always come first, we had no more problems.

By September, Chewy was a large animal. The hump was well developed, the bell beard hung below his jaw, the bumps of developing antlers began to appear. He was restless; the enclosure was not big enough. He paced around and around the edge looking out into the wilderness which surrounded it — available to him — except that it wasn't. Not until after hunting season. His colour had changed from the brilliant bronze of the calf, to the gleaming chestnut of a full grown bull; out in the wild, he would be beautiful. By the end of November, he was so big that I was looking up at him.

We had to wait. We were warned by some of our local friends that a couple of macho young men, hunters, were just waiting for his release; somewhat accustomed to humans, he would be an easy target. Surely even they would respect the end of the hunting season? Wouldn't they? Anyway, the official hunting season ended.

The decision to release him was difficult. Winter was coming. He did not like to eat grain — he wanted to browse. I think every willow tree along the back roads was shuddering when it heard the approach of our truck — but we could no longer supply the amount of browse he wanted. We hoped he would settle for the acres of tag alder and willow and tamarack behind the Sanctuary along with the relative safety there — but we had to release him.

So, one blue and golden morning early in December, the Sanctuary assistant took down part of the fence and we stood back to watch Chewy go free. For several moments, seeming to consider, the moose stood looking at the gap in the fence. Then he stopped and took a long, cool drink from the pond. Again, he moved slowly to the gap and stopped again, looking out. What was going through his mind? Curiosity? Apprehension? Wonder? Still slowly, he walked out of the pen and down towards the back fields, the rim of tag alder, the miles of swampy lands — always slowly, as though he were thinking at every step. Sometimes he stopped, head up, looking over the fields. Down, amongst the bushes, he moved more quickly. He looked toward us, once. Briefly, he moved toward the Sanctuary assistant — and then, suddenly, he began to trot. Head up. Quickly.

He vanished amidst the tag alders and the tamaracks. For a moment we heard him and then we didn't. Stillness. Chinneas had gone free.

Once again a large animal had been released into the bush, and once again I was expecting a long, quiet night. I would think about Chewy. I would wonder how he was enjoying finding all the willow and alder for himself. Did he fully realize that now no fence surrounded him — I wondered, and long after the night closed in, I fell asleep.

And awoke to the sound of a truck cruising slowly along the road with a strong light searching each side of the road, shining deep into the woods. Jacklighters. Illegal. Lacking any trace of sportsmanship — but happening. Word must have gone out that the moose had been released. I pulled a bathrobe around me, stuck my feet into my boots and went out — I had a flashlight, too — not as strong as the hunters used to probe the woods, but all I wanted was a license number. However, I am not very skilled at law enforcement. The hunters saw me, gunned the truck past so closely and so fast that I did not get the number, and vanished around the bend and up the hill.

But they never came back.

Winter closed in; the deep snows arrived in December, and in January, the bitter cold. The bears were asleep, the raccoons were fat and slept most of the time. I had time to go walking with my dogs, up along where the drifts were not deep. The road wound through the open valley and up through the woods. I tried to keep a trail open along the top of that high hill which we wistfully call "the mountain". From there we can see for miles over the valley fields, over the wetlands, and then to the woods beyond.

One day, dressed warmly against the February wind, my dog, Abby, and I were walking slowly up the road. Abby wasn't really enthusiastic about being out walking in the cold, but was willing to go along with me if I had a mind to go. I glanced across the meadow just below the mountain and saw Chewy, nonchalantly chewing his cud, unalarmed by our appearance. I spoke his name, but, totally ignoring us, he continued chewing. Not really keen on encountering a full-grown bull moose, who just might have learned how to be wild, I decided the wisest course was to ignore him, too. The next day, from the high point of the mountain, I saw him lying in a small clearing he had trampled out in the alders. For a while, I watched him. I saw the broken branches through the swamp, where he had been eating. But again, I did not call him.

Whether he did not like being ignored or whether seeing us had triggered something in his memory which suggested that we would be a good source of food — who knows? A few days later, as I was returning from feeding the birds in the barn, I saw Chewy strolling casually in between the cages and coming to a stop in the lane — between me and the house; I stopped. The only animal which had ever seriously injured me was a buck deer, a midget compared to the huge bull moose who stood regarding me. May I be forgiven for feeling some apprehension . . . and for stooping to bribery. I returned to the barn, brought out a dozen apples and scattered them onto the ground. Kneeling, as moose do, to eat them, Chewy was content, and I got into the house.

Those apples set the platform for the remainder of the winter. Once a day he would come strolling up from the swamp. Once a day I would roll apples out onto the snow. And he would kneel to eat them. (I took several pictures of the kneeling moose — again, I shouldn't wonder, that ridiculous position an example of God's sense of humour; neck so short he couldn't reach the ground, massive rump lifted to the skies. I wanted to send a picture to some religious magazine and label it: "Moose at Prayer".)

I had a porch, surrounded by a wall, and a half-door. Chewy quickly learned that, if I did not serve the apples immediately, he could come to the door and stand with his head over it, staring into the house. If I still didn't come immediately, he would ring a set of lovely, tuned chimes which hung in the doorway.

Gradually, wisely or not, I began to trust him. Somewhat. He took to staying in the lane. After he had been fed, he would follow me up to the barn and wait until I came out again. I still had to walk the dogs but usually I managed that in the daylight. One night, when the cold was not too bad, and the great white flakes of snow were falling thick and fast, I wanted to walk. I looked out, peering through the whirling snow and the darkness. No sign of Chewy. Abby on a leash, we went down the lane, and turned to walk up the road. We had gone about a hundred feet when, not warned by the slightest sound, I felt a huge nose on my shoulder, and moose breath in my ear. Very quietly, Abby and I turned around and retraced our steps up the lane and into the house.

Chewy remained with us all that first winter. His only other encounter was with those Eastern Timber wolves. Full grown Eastern Timber wolves are not large and these were still less than a year old. The dominant female, Winter Moon, was convinced she owned the world — and though the enclosure fence was over ten feet high, she could not endure the sight of a young moose walking free. Some instinct inside her told her that she was a wolf, she could take a moose any time — and so, once, as Chewy was strolling by the wolf enclosure, Winter Moon sprang at the fence. Much to her surprise, and ours, and Chewy's, she was over the fence. And chasing him.

He ran. She chased him almost joyously. Then it must have occurred to him that that was a very small wolf. He turned and chased the wolf — and, as she ran, it must have occurred to her that wolves chased moose, moose did not chase wolves. She turned and chased him. How long the situation would have continued, I do not know. The Sanctuary assistant, who had been caring for the wolves, stepped between them, caught the wolf and tossed her back over the fence. Dazed, she stood and shook herself. She never managed to figure out how she climbed the fence that once.

When spring finally came, Chewy vanished into the woods. When food was plentiful, moose need neither apples nor humans. We saw no sign of him until the following winter, when, once again, bigger, with antlers, he came strolling up the lane. I was glad to see that he had made it through the summer, that he had survived the hunting season, but at the same time, I determined that I would not feed him. Except that he kept ringing the chimes.

Nor would he let me out of the house until I fed him. On Christmas Day we had a standoff.

I was invited out for Christmas dinner. Between me and my car stood, firmly, one huge bull moose, head lowered, staring at me. Finally, I tossed the apples over to the other side of the car and made a run for it.

We try not to let the animals we raise become too tame; and the bull moose always has the potential for danger. They know humans apart, and may choose to be a friend of one and not of the other. When the head is lowered and the ears laid back, be careful! Very careful. The Sanctuary assistant had known Chewy all his life. One day, that second winter, he was working with his chainsaw away back in the swamp. His snowmobile was beside him. Likely the noise of the chainsaw prevented him from hearing the approach of a very agitated moose. Chewy erupted from the woods, antlers swinging, hoofs high. He plunged into the snowmobile, ripped the hood from the motor — which caught between the antlers. He swung it back and forth and back and forth, until it flew free. Then he turned on the assistant. He, in turn, started the chainsaw, threatening, hoping the noise would drive Chewy away.

The saw stalled. The only weapon remaining for the assistant, was his snowshoe — which he swung at the moose — the snowshoe broke.

The assistant tried to head for home, keeping the small trees between him and the raging moose — an interesting but dangerous dance. He managed some distance before he heard the sound of an approaching snowmobile. Chewy heard it, too, and fled.

Experiencing, with a wild animal, a relationship of mutual trust, is a profound privilege. If I said, "some of my best friends are beavers", I would sound frivolous, perhaps, in some people's estimation, somewhat silly. Nevertheless, some of my best friends are beavers. I have had the experience of walking alone and quietly through the woods, until gradually aware that I was not alone — a raccoon walking beside me, or a deer. I remember sitting on a log at the edge of a meadow, and a hawk which I had cared for and released the previous year, circled down out of the sky, its great wings spread, and then folded as it came to rest nearby — stayed awhile — and then spread its wings and returned to the high skies.

Though, since that first experience with Chewy, I have, over the years, known a good number of moose calves, watched them grow to adulthood and go free in the bush, I have never known real trust. Perhaps that is because they sense my apprehension. Tony, presently the manager of the Sanctuary, is now in charge of moose; he has moose friends. Havoc, a moose whose name is appropriate, was raised from a tiny calf. Now, free, she will greet Tony when they meet in the woods. She will follow him back to the Sanctuary, where, if she meets me, she will lower her head and charge. Therefore, I do not blame Chewy that we were not the very best of friends. I just wish Tony had been around back then . . . .

Chewy did not dislike humans, which was unfortunate; humans are the worst of moose predators. He didn't especially like them, either. But he was curious. Early one winter, when the hunting seasons were over, I happened to be talking to a young man who had been out hunting for deer. He said: " I was just sitting on the trail waiting and then this big moose come up behind me — a big fella — he nosed at my hat and gun — then he went on up the trail. You can't shoot a guy like that. Besides, moose season is over."

Another group of young men, some miles from here, were working clearing a field, when a moose ambled out of the woods to watch them. When he approached more closely, they encouraged him. Later the next week the village was entertained by photographs of the men with their arms around the moose. Another man, working at the edge of a field with a front-end loader, saw a bull moose approaching him. Knowing, rightly, that a bull moose in the fall can be dangerous, and failing to frighten the moose away, he climbed into the bucket and raised it as high as he could. He said that the moose came, tried to nose the bucket, so he leaned out and patted the moose.

These stories, told to me because people thought that I would be glad that Chewy was doing well (I was!), nevertheless, raised real apprehension. I knew all humans would not be so amused.

Chewy returned to the Sanctuary with an antler splintered off. I wondered if someone had taken a shot at him and narrowly missed.

We knew he had survived another hunting season when he showed up at the beautiful log home of neighbours down near the end of our road. They had a lovely porch, complete with a hot tub. Friends from Toronto were visiting — they were having a very, very good time in the hot tub — laughing, noisy. Then Sandy saw, over the shoulders of the others, a big moose advancing up the driveway. Saying absolutely nothing, he watched as Chewy ambled closer and closer. Stopped. Looked. Came over. Raised his head over the edge of the porch and probed the nearest man with his nose. A story to take back to Toronto.

Winter came. The cold. The deep snow. As seemed to be his custom, Chewy disappeared into the bush. We never saw him again.

Winter Moon, an eastern timber wolf.

Epilogue

The wild will always have in its innermost being — Mystery. No matter how our wilderness is invaded by researchers, biologists and writers — the wild will answer us, wondering at our intrusiveness. The fundamental nature of each creature is full of secrets which we cannot know.

Some are, perhaps, a little amused at our curiosity.

Yet we do not, we cannot, conquer our compelling inquisitiveness. We want to explore. We want to understand as much as we can. And, it is vastly important that, to the best of our abilities, we do try to know. Though we may not have the ability to understand completely, unfortunately we do have the capability to destroy completely.

We should, and must, learn all we can.

We must learn to respect each creature. The value of the wild is far beyond mere economics, far beyond what we consider our rights and traditions. We must accept what history can teach us; many traditions (the slave trade, torture, trapping, war, hunting) are not sacred in themselves. Life is beyond value. We must learn to experience the joy of the wild without destroying it.

I have known the companionship of wild things a privilege beyond mere description, an experience for which, perhaps, we were all created.

Sanctuary trail.

Aspen Valley Wildlife Sanctuary

Sanctuary: "A place of refuge" says the dictionary. In the case of the Aspen Valley Wildlife Sanctuary that is indeed an apt description. This Sanctuary is a place where orphaned and injured wildlife are taken to be cared for, some so small, some so thin, that they must be fed with a syringe, or so injured that they require similar care to that which humans would receive in a hospital intensive care unit. In one winter recently, close to 100 bear cubs were cared for because their mothers had been shot during the fall hunting season. Bears, coyotes, wolves. beavers, deer, moose and other animals regularly find food, shelter, and care at AVWS. When the time comes and the creatures no longer need the care of the Sanctuary personnel, they are taken to an appropriate location and released back into the freedom of the wild.

Fifty years ago there was no thought of such a Sanctuary near Rosseau in Muskoka. It happened almost by accident. Audrey Tournay was teaching High School in Parry Sound and decided that she would like to live on what used to be a farm. The property she found had a comfortable small house suitable for her and her dogs, a large, sturdy barn which could house her two horses, goats and a donkey along with her domestic animals, and lots of outdoor space to roam and explore.

However, Audrey had always been interested in wildlife. There was plenty of that around the farm property. She became known as a person who was willing to care for orphaned and injured creatures. Gradually neighbors, then the Ministry of Natural Resources would bring to her such animals for care.

Over the years the Sanctuary has grown so that now it has acquired over 1000 acres of land, has a Board of Directors, a full-time staff of 5, and hundreds of volunteers — mostly from various parts of Ontario and the rest of Canada, but also from such countries as Germany (which sends groups of students each year to help and to learn about the care of wildlife),

Now AVWS is recognized as one of the most outstanding Wildlife Sanctuaries in the world. Recently Audrey and Tony Grant, the Sanctuary manager, were joint recipients of an honour awarded by The International Fund for Animal Welfare with an Animal Action Award. Tony has also been recognized as a world specialist in the care of bears and was invited recently to travel to Russia (with all his expenses paid), to attend a conference on bears and to speak of his experiences about the care of bears.

The education of others, especially children, has always been important to the staff of the Sanctuary. Today the property has its own schoolhouse where classes from nearby schools come to learn the wonders of wild creatures. The purpose of this educational enterprise is that the children will know and learn to respect the wildlife around them.

The total operation of the Sanctuary has grown to have a large cost. The staff provide opportunities for visitors to learn of Sanctuary activities by having visitor days and a few special events where participants may make contributions toward operational costs. However the largest amount of help comes from many supporters who contribute regularly with small or larger financial gifts. The Sanctuary is now registered as an officially recognized charity which is able to provide income tax receipts.

Thus, from small beginnings Aspen Valley Wildlife Sanctuary has grown to be a well-recognized organization for Wildlife conservation in Canada. However, though recognition is important and appreciated, the focus of all that is done there is still the welfare, care, and rehabilitation of the wild creatures who are brought to the Sanctuary for sustenance, healing and compassionate care.

— Written by a friend of the Sanctuary

On the Edge of the Wild